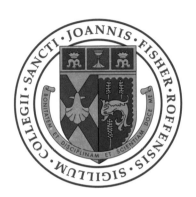

THE RETURN OF THE WOLF

THE RETURN

OF THE WOLF

Reflections on the Future

of Wolves in the Northeast

by Bill McKibben, John B. Theberge, Kristin DeBoer, and Rick Bass

John Elder, Editor

Middlebury College Press

PUBLISHED BY UNIVERSITY PRESS OF NEW ENGLAND

HANOVER AND LONDON

Middlebury College Press

Published by University Press of New England, Hanover, NH 03755

© 2000 by the President and Trustees of Middlebury College

Pencil drawings by Mary Theberge. Used by permission.

Printed in the United States of America

5 4 3 2 1

Library of Congress Cataloging-in-Publication Data

    The return of the wolf : reflections on the future of wolves in the

    Northeast / by Bill McKibben . . . [et al.] ; John Elder, editor.

        p.   cm. — (Middlebury bicentennial series in environmental studies)

        Includes bibliographical references.

        ISBN 0–87451–967–5 (cloth : alk. paper)

        1. Wolves — Reintroduction — Northeastern States. I. McKibben, Bill.

    II. Elder, John, 1947–   III. Series.

QL737.C22 R475 2000

333.95′9773′09741—dc21   00–010463

Contents

THE RETURN OF THE WOLF

Introduction

SCIENTISTS, POLITICIANS, CONSERVATION GROUPS, AND public agencies have all begun looking at the northeastern United States as a region where packs of wolves might once more be able to live. It is remarkable to find such a conversation taking place in this long-settled and populous part of the country, where the last wild wolf was killed well over a century ago. Public controversy has followed hard on the heels of such speculation, with bills to prohibit the reintroduction of wolves already debated in the state legislatures of Maine, New Hampshire, and Vermont. In the scientific realm, too, questions have arisen about whether restoration projects are called for, as well as whether "recovery" accurately describes the genetic morphing that occurs when wolves independently expand their range and interbreed with coyotes.

If the future of wolves in the Northeast remains uncertain, however, we can at least describe some of the forces behind the current dialogue. One of the most striking stories of our century, both scientifically and culturally, has been the shift in attitudes toward predators. It's important not to overstate this change. Bounties on wolves still exist in some places, and the debate in New Hampshire showed that plenty of people hate and fear these animals despite having had no contact with them. Nevertheless, the science of ecology has gone a long way toward educating the mainstream of our society about the complex balance necessary for a healthy natural environment. The classic description of wolves' significance for a landscape is Aldo Leopold's "Thinking Like a Mountain," from *A Sand County Almanac* (1949). In that essay, he tells of shooting into a pack of wolves when he was a young man in New Mexico territory, and of discovering a new perspective on conservation in the green fire of a dying wolf's eyes. Leopold became an early and influential proponent of predators, understanding that "a buck pulled down by wolves can be replaced in two or three years, [while] a range pulled down by too many deer may fail of replacement in as many decades."

The reintroduction of wolves into Montana, New Mexico, North Carolina, and the Great Lakes region over the past decade represents a culmination of Leopold's insights, as well as of the intervening decades of research by wolf biologists and ecologists. The Yellowstone initiative was particularly controversial, with ranchers fearing decimation of their stock by the reintroduced predators, antagonistic legislation quickly being proposed, and highly publicized threats to shoot any wolves that wandered over the line from the National Park. While the results are still not all in, the restoration project in Yellowstone seems, in its initial years, to be going extremely well. Agreements have been forged to compensate ranchers who lose animals to the wolves. Communities on the edge of the Park have experienced an economic boom from wolf-related tourism, with some former opponents of restoration now deriving part of their livelihood from guiding those who want to see the new packs. And the wolves themselves have begun to establish themselves in their new environment, reproducing at a good rate and bringing enhanced elasticity and balance to the populations of elks and other animals already in the landscape.

It's easier for us to imagine the howl of wolves in the grandeur and vastness of Yellowstone than amid the villages of northern New England and upstate New York. The fact that we are beginning to do so, however, relates to a second change that has occurred over the past century. Trees, as well as ecological awareness, have been growing in this region. Bill McKibben has written of "an explosion of green" here. Vermont typifies this forest history, having gone from being about 80 percent deforested at the time of the Civil War to being almost 80 percent reforested today. Although the Northeast as a whole is densely populated, the northern tier of it—increasingly referred to as the Northern Forest—now encompasses expanses of densely wooded land far greater in acreage than Yellowstone, Glacier, or Yosemite. Bear and moose have already made a dramatic comeback in many parts of the Northeast, with signs that catamount might possibly become re-established, too. This is the context in which wild corridors down from Canada are now being mapped—zones along which wolves might be restored, or might return on their own, to the Adirondacks or to Maine. Northern Vermont and New Hampshire are being assessed as landscapes into which the re-established packs in those states might naturally expand.

The four essays in the present volume respond to the exciting, if uncertain, prospect of wolves returning to the Northeast. They comprise neither a comprehensive report nor a unified argument for any given policy. Rather, in accord with the bioregional approach of this Middlebury College/University Press of New England Bicentennial Series on Environmental Studies, they reflect upon biological and political realities and their cultural implications. In the opening essay, Bill McKibben asks why such interest should arise now in an animal that our society formerly took such pains to exterminate. Is this just the latest, environmentally cloaked, version of consumerism? If the desire to restore wolves here turns out to be more significant than that, what changes might it require in our own human population and habitat, as well as in those of the wolves? John Theberge draws on his years of ecological research among the Canadian populations from which any returning northeastern wolves would, one way or another, be drawn. But he is careful to point out that the "return" of long-absent animals is much more genetically and behaviorally complex than either the boosters or the opponents of such a process might assume. Kristin DeBoer writes as a wolf-advocate working for RESTORE: The North Woods, the first conservation group to advocate for restoring wolves to the Northeast. She both describes the current state of the political struggle and evokes the spiritual and emotional influences that have led her to dedicate herself to this work. Finally, Rick Bass looks at the debate in our region from his own background as a writer and activist in Montana. Like McKibben, he finds cultural questions at the heart of what might be seen more narrowly as a biological process. He argues that, unless wolf recovery is connected both to preservation of wilderness and to a more economically and culturally sustainable model of forestry, the mere presence of wolves will offer little lasting benefit to our region.

The debate over the return of the northeastern wolf—and the wolves' own feints and forays in the Northern Forest—will continue to be played out in coming years. The four essays here contribute their own distinctive and thoughtful reflections to this process. They remind us that the future of our entire region is connected, in vital and surprising ways, to the drama of these gray shadows just at the edges of our field of vision.

Human Restoration

Bill McKibben

THE HILTON HOTEL ON THE SLOPE BELOW THE STATE capitol in Albany, New York, houses an endless rotation of people who would like the Empire State to behave in some particular way. One day it's the snack food dealers demanding lower taxes on potato chips, the next it's landlords fulminating against rent control, the next it's paving contractors with lists of new roads to build. They pass through in groups, distinguishable by name tag, feeding on buffets and open bars and trays of shrimp; when they meet a legislator they either shyly flatter or browbeat him depending on his seniority or vulnerability. They press the flesh, rub shoulders, build new relationships or re-establish old ones, elect leaders, and then they migrate on, back to the daily grind of selling Fritos or bitumen.

I've seen timber wolves exactly once in my life—a pair of them, in the downstairs ballroom of this very Hilton in the fall of 1996. An environmental group called Defenders of Wildlife was sponsoring a wolf conference at the hotel, with experts from all over the nation. "We chose Albany as our conference site in order to draw attention to our proposal to restore the eastern timber wolf to the Great Woods of the Northeast," wrote the president of Defenders. In other words, they wanted something from the government too: permission to stick wolves back into, say, the Adirondack

Park in northern New York, where I live, or the Maine Woods, or the White Mountain National Forest, or Vermont's Northeast Kingdom. Even though they were more earnest than the snack dealers, the Defenders came with the same sorts of apparatus—charts, polling surveys, the implicit promise that there might be a political payoff for those who supported such works. And they came with props—a pair of captive-born wolves from a wolf center somewhere out West. The conference organizers were savvy enough to know that two or three reporters might cover a conference full of papers on topics like "Evaluating Vasectomy as a Means of Wolf Control," and "Tracing the Life History of a Dispersing Female Wolf in Northern Wisconsin," but that every TV station in the neighborhood could be counted on to send a camera if there were living wolves to shoot.

Some functionary ushered all the reporters and camera crews into the subterranean ballroom; everyone plugged in their mikes; speeches were heard and literature distributed. And then, finally, a pair of handlers ushered in the wolves themselves. And suddenly everything changed. A cold front blew in, cutting the damp human-flavored air. The wolves showed next to no interest in the dais full of whirring motordrives and humming videocams; though they were on leashes, the room was theirs by right and they explored it thoroughly. One saw a thermostat high on a wall, and lifted up on his hind legs to look at it, stretching his long body till his muzzle reached the dial. The other stretched out on the carpet a moment. Those deep unselfconscious stretches, almost more feline than canine, declared them masters of the terrain—none of the tense jostling of photographers eager for a tighter shot, none of the favor-begging of the environmentalists. They were simply there. Not larger than life. It's just that human life seemed a bit smaller in their presence. They were life-size. They fit their skins. They didn't charge the air. They relaxed the air.

Since I've now exhausted my firsthand knowledge of wolves, let me move on to a topic I know something more about: a forest without wolves. I've spent some time in the New England forests, but my real experience is confined to the Adirondacks, a range of mountains that bulges out in a great dome from northern New York. Bordered roughly by Lake George and Lake Champlain on the east, Lake Ontario on the west, the St. Lawrence on the north, and the Mohawk valley on the south, the Adirondack Park covers a swath of land larger than Glacier, Grand Canyon,

Yellowstone, and Yosemite national parks combined. On a map of the eastern U.S., it's the large patch of green in the upper right hand corner; if you fly over it in a light plane, that green patch on the map becomes a green patch of ground—unbroken trees as far as the eye can see, none of the clearcuts that checkerboard Oregon or Washington or Maine. It's just trees and mountains and lakes and rivers, interspersed with small towns scattered every ten or twenty miles.

What makes it interesting, however, is less its current beauty than its scarred past. Remember these two dates: 1899 and 1907. In 1899, St. Lawrence County paid the last bounty in northern New York on a dead wolf; as far as anyone knows, that was the last *Canis lupus* to lope through these mountains. Her death marked the symbolic end of an era of spasmodic destruction that had begun in the early nineteenth century, a space of six or seven decades that saw a nearly primeval forest cut over, burned, plowed, and polluted with great ferocity. People had come late to the central Adirondacks—the Indians used the mountains only as a seasonal hunting ground, and the first European didn't climb the range's highest peak until 1837, a generation after Lewis and Clark had returned from the Pacific. But when they came, they came with a vengeance. The first images of environmental destruction chiseled into the national imagination were lithographs and early photos of bare Adirondack hillsides sliding into creeks, of moonscape vistas from high peaks.

The epic destruction coincided with the birth of what we'd now call adventure tourism, as city dwellers rode the rails to escape the Industrial Revolution for a few weeks and enjoy this accessible paradise. (The first public building on Earth with electric light was the big hotel on Blue Mountain Lake; that should give you some sense of it.) These tourists helped spur the efforts to protect the forest, which also gained support from downstate tycoons worrying that deforestation would silt in the economically all-important Hudson River, which rises in these mountains (an insight they gained at least in part by reading best-selling proto-ecologist George Perkins Marsh, the Vermont polymath who was then midwifing environmental science). Together they passed landmark laws creating the forest preserve, and began the slow task of buying land within that six-million acre park. It's murky work trying to fix the right date for the end of the park's decline—while the state constitution was amended to draw a "blue line" around the Airondacks in the 1890, the most destructive logging may not have taken place until the early years of this century. But 1899 will do—the last wolf howl strangled, the wildness of the place seemingly gone. No bears were left, or almost none; no mountain lions, elk, or lynx. The Adirondacks had been subdued.

And then, eight years later, in 1907, something happened that seemed unimportant at the time. State officials released a few pairs of beavers from Yellowstone Park in a couple of drainages in the southernmost part of the Adirondack Mountains. The beaver had been extirpated too—trapped out for its pelt, its habitat cut down, dried up, dammed for mills. But now a few people were interested in bringing the beaver back. And it took. With protection from game wardens, the animals started to spread north, back into forests that were slowly recovering—many of them forests the state had bought, or taken over for nonpayment of taxes, forests protected as "forever wild" under the state constitution, where no human could cut a tree. The law did not apply to beaver, of course, and before long the early succession white birch were lying on the ground beside one pond after another, before long the dams were backing up trickles into wetlands on a thousand watersheds, before long the tailslaps of nighttime warning were cracking out across three thousand ponds.

This was the beginning of Adirondack restoration, one of the earliest such efforts anywhere. From 1907 on, people were trying not just to protect what was left of the Adirondacks, but to recover some of their original character. A great experiment was underway: could you unscramble an egg? Bruce Babbitt, Bill Clinton's gregarious Secretary of the Interior, has insisted in recent years that restoration must be the hallmark of twenty-first-century conservation; he's already helped take out a few big dams. But for the Adirondacks (and in a subtler way for much of the eastern forest), the twentieth century has already been a restoration epoch. So what news do we have to share?

When I think about restoration, as I said, I begin by thinking about beavers. I try to imagine what these woods would be like if someone had not trucked the beavers out from Wyoming and let them go in a couple of streams. It would be much quieter, I think—the noisiest places I know in the woods are murky small sunlit ponds backed up by some intricate dam or another, algal places filled all summer with every kind of buzzing and droning and croaking. The beaver waddled back into the Adirondacks trailing life of infinite variety behind him. That's the first thing, obviously; restoring animals means restoring processes.

Other creatures wandered back too. I know old men in my town who can remember the first bear they ever saw, sometime in the 1930s. Now there are trees enough to hide them by the thousands: they hoot from the ridges by night, they leave great piles of berry-colored scat on the trail and wide clawtracks on the smooth-skinned beech. Every once in a while they get so lost in their eating that they don't hear you coming and then they finally do and crash off through the undergrowth, making your heart beat hard. I can't imagine this landscape without bears; it would seem *lean*, austere. Restoring animals means restoring the spirit of a place.

As the animals have returned, the forest has returned; ghost flora conjuring up ghost fauna and vice versa. About half the Adirondacks now belongs to the state as inviolate wilderness; the other half is in private hands, used mostly for timbering, but under fairly tight forestry laws that prevent the abuses common elsewhere. There is enough land that the moose wandered back about a decade ago, somehow crossing the highways and lakes that separate us from New England.

People reacted to the return of the moose in an interesting way. When big environmental groups, with tacit support from the state, proposed reintroducing the huge animals, many local residents hated the idea. In public meetings they worried aloud that the moose might spread disease to the deer herd, and that Adirondackers would surely die in collisions with the beasts (the haunch of a moose is just about windshield-high). And so the state backed off. But just about the same time, the moose began reestablishing themselves, and no one complained at all about that. In fact, people loved it—when a moose is spotted in our town, the phones begin to ring as everyone shares the news. Nature was simply taking its course, and that didn't seem to bother anyone.

If the wolf would simply walk back in—if a pack from Algonquin Park in Ontario would simply decide to wander south a few hundred miles—then I suspect they too would be greeted with open arms. But that isn't going to happen, because big highways, broad stretches of open farm valley, and the St. Lawrence River all stand in the way. The wolf, killed off by people a century ago, needs people to bring it back. Which is a problem.

Even two decades ago, the idea of reintroducing wolves here would have been a non-starter, ludicrous. But it's easy to underestimate how successfully environmentalists have shifted people's thinking, at least about animals. The Red Riding Hood crowd is smaller now, and mostly older. At least as many people have some Farley Mowat story stashed away in their hearts. When Defenders of Wildlife commissioned a poll in 1996, a huge majority of New Yorkers, and a substantial majority of Adirondackers, favored bringing back the wolf.

That doesn't mean a restoration would happen easily, though. Among the park's political powers, there is little support for reintroduction. County governments have passed symbolic resolutions prohibiting the release of "dangerous predators" within their borders; various barroom heroes have boasted to the newspapers that they would kill the creatures if they ever appeared. Some of their arguments have been pretty tenuous; a great deal of concern is expressed for the almost nonexistent livestock of these forested mountains, for example. The visceral opposition comes mostly from people who don't want to be pushed around from outside, who don't want someone else telling them what to do. In the words of a leader of the Adirondack Association of Towns and Villages, approached by a reporter on the day that Defenders first suggested a reintroduction: "I don't know much about it, but I'm sure we're against it." Against some outside notion of what constitutes the proper forest.

No one even knows if a reintroduction would be successful—the first reports from biologists cast doubt on the ability of the Adirondacks to support a pack, in part because they fear too many would be shot. But the deeper debate goes on outside the facts. What constitutes a wilderness restored? Is the process begun in 1907 still underway, or has it reached its limit? How far can restoration go in a place where human beings live? Because, of course, what makes the Adirondacks interesting, besides its scarred past, is its peopled present. This is not like Yellowstone, where you need to worry about the surrounding ranchers; here, wilderness and human settlement are completely intertwined.

We know that human beings can wipe out most other large creatures; we know that if humans set aside a large enough wilderness they can save most creatures (though we're only beginning to appreciate how large, and how intact, that land must sometimes be). What we don't know—and it's a fairly important question for a world of six billion people headed for ten —is whether in the modern age you can have small but complete human settlements amidst intact wilderness. That's the experiment the Adirondacks, without exactly knowing it, undertook a century ago; that's the experiment that environmentalists have begun in many tropical regions within the last decade. It's still a new experiment. The results are only starting to trickle in.

Often, in the winter, I'll be out skiing on some backcountry lake and come across the stomach, the head, the hooves, and some of the skin of a deer. That's all that's left a day after the coyotes have chased it out on the ice, the whitetail's skinny legs windmilling till they snap. When the coyotes eat their fill, the ravens descend, and so on down a long cafeteria line.

One way of asking if a place has been restored is to ask if its biology works, if something is missing that causes something else to be out of balance, and so on. If you visit my in-laws' house in suburban Connecticut, the answer couldn't be clearer. They have no rosebushes left. Pie plates dangle from their shrubberies in an effort to keep the deer at bay. Does and fawns browse with utter unconcern in the backyard; if you run out to chase them away, they look at you with unconcealed contempt. Since you can't hunt very well in this land of cul-de-sacs, and since there are no predators, and since people have created a smorgasbord of browse with their lawns and gardens, it is deer heaven. As Richard Nelson makes clear in his classic book *Heart and Blood: Living with Deer in America*, most of America is now deer heaven for just such reasons: "By 1991 an estimated 1,500 whitetails lived within the city limits of Philadelphia. That is more than the total number of deer inhabiting the state of Pennsylvania in 1900."

But in the Adirondacks you don't see so many deer. They are there—on many lakes you can see a good browse line, where they've nibbled the lower branches of the shoreline hemlocks to the same height as they stood on the ice—but three things have kept their numbers in some kind of check. First, there's not a lot of opening in the forest; without clearcuts, and with three million acres of the park off limits to any logging at all, sunlight doesn't reach the forest floor in sufficient quantity to grow great pastures of browse. Second, humans hunt them. And third, in these mountains the coyote has grown into something like a wolf.

Coyotes are supposed to be lone scroungers, loping through the world looking for an easy meal. They are superbly adaptable (a biologist I knew once said, "when the last man dies there will be a coyote howling over his grave"), but we don't think of them as a top predator. In the absence of the wolf, however, that may be how they have evolved in parts of the American East. Our coyotes are big, they have learned to hunt in packs, and they eat a lot of whitetail—the 1,200 coyotes now living in the Adirondack may consume 12,000 deer annually. So if you have a wolf substitute, why do you need a wolf, especially if it might mean Government Restrictions of some kind?

The biological answer is not completely clear. Wolves and coyotes aren't identical; coyotes haven't learned how to kill beaver as efficiently as wolves do, for instance, and so beaver numbers are booming (just ask the road crews who spend half the year draining flooded roads). And despite the fears of the sportsmen, wolves might actually benefit human hunters; if the biologists are right, they will chase off lots of coyotes, perhaps enough to actually diminish the wild deer kill. But it's not an overpowering argument. The mountains like wolves, as Aldo Leopold pointed out so memorably, but they have probably come to appreciate the coyote as well.

There's a kind of theological argument too, of course. Wolves were here, and we chased them out, so we should invite them back. Amen, hallelujah, thanks be to Noah.

I've felt the power of this kind of reasoning. A few years back I spent some time on the Alligator River National Wildlife Refuge on the Atlantic shore of North Carolina, an aggressively nasty place. Old pine plantation, absolutely flat, gridded with dirt roads that ran at right angles, microwaved by the Sun, guarded by nine kinds of things that fly and bite, and two or three that slither and bite. I'm sure that if you grew up there you might develop some slight affection for the place, but every tourist I saw was heading swiftly for the Barrier Islands. Yet something incredible hung in the air, for it was in this place that *Canis rufus*, the red wolf, had been resurrected.

Red wolves once ran across much of the continent; they may have made it up into the Adirondacks. But by the 1980s, just a few were left in the United States, hanging out in Texas. They weren't going to hang on many years more, so federal officials hauled the last ones into zoos. This strand of DNA, this voice on the land, this particular fancy of creation, seemed to have run its course, and not just in one place, like the Adirondack beaver, but every place. Forever. But the environmentalists and the feds hadn't quite given up, and they found a piece of land in eastern North Carolina, in the former range of the red wolf, to reintroduce them into the world. In 1987, the Fish and Wildlife Service opened some cages and released four pairs of captive-born adult red wolves. By the mid-1990s, ninety-four pups had been born in the wild. These were fourth- and fifth-generation wild wolves.

I didn't see them—I only see the wolves in hotel ballrooms. A Fish and Wildlife worker drove me around the refuge for long hot hours, sticking rank bait in traps as he tried to catch some study animals that needed new batteries in their radio collars. I didn't hear them howl or bark; I just heard their beeping transmitters as I held the antenna out the window and we tried to home in on them. But that didn't much matter—it was the simple fact that these wolves existed that mattered. For a while the red wolf had not existed; it was alive in a zoo somewhere, but that was nothing more than keeping its DNA on ice. It hadn't existed as part of the sweet battle and concert of life bumping up against other life, and now it did again. God smiled.

For the timber wolf in the North Woods, though, the case is not so clear. Wolves live other places—lots of them in Alaska and Canada, healthy and well-established populations in Minnesota and Wisconsin, the new packs roaming Yellowstone. They have become accepted in most of those places, just as the red wolf has become accepted in North Carolina. Long caravans of cars go out on nighttime wolf-howling expeditions. In Minnesota, they named their damn basketball team for them, and tens of thousands of people visit the International Wolf Center in Ely, a hard-to-get-to little town hard by the Canadian border that now boasts daily plane service from the Twin Cities. At least on this continent, wolves seem safe as a species.

So if the wolf doesn't need to be here, and the deer don't need them here, who needs them? Might it be the same species that needs cable TV, high-speed modems, hair transplants, twenty-one-speed mountain bikes, frappucinos, frappucino grandes, frappucino tres grandes? Could it be the same species that needs two-thousand-square-foot houses and monster trucks and 17-gigabyte hard drives? Could be.

The world as we now know it is shaped by human desire, especially by the incredible needs developed by the affluent West in a single century. Those needs shaped the landcape in obvious ways—the suburbs with their endless circling roads are desire poured from the back of a cement mixer, the desire for convenience, comfort, privacy, individuality. And now those needs shape the landscape in more insidious and permanent ways. You can read our desires, our needs, in every cubic meter of the air, and in every upward inching of the thermometer. Our needs for mobility, for elbow room, for never getting too hot or cold or bored, all require the consumption of gas and oil and coal, and that endless combustion fills the atmosphere with carbon dioxide. Spring comes to the northern hemisphere a week earlier now than it did two decades ago, because of our needs. The warmer air holds more water vapor, and hence severe storms are twenty percent more common—that's our appetite at work. The warmer oceans expand, and glacier melt pours more water into the sea; the scientists say that the Atlantic and the Pacific will be a couple of feet higher before this century is out, thanks to our needs. We are machines for generating needs and then generating technologies to fill those needs (and sometimes the other way around). So what might it mean to say we need the wolf?

We need something new to consume. We need wolf howl, we need the little frisson that comes with thinking: there's a wolf pack out here in these woods where I'm spending the summer. This kind of symbolic consumption is old news, of course. Folks buy Durangos or Blazers or Pathfinders or Dakotas or Tahoes or Expeditions or the other behemoths of the sport utility tribe knowing full well they'll never take them off road. The extra fifty horsepower and thousand pounds of steel and eight inches of clearance give them rights to a certain image. You can buy it for two bucks in a pack of Marlboros. In the same way, I'd like wolves in my backyard because I'd like to think of my backyard as wonderfully rugged. The more rugged my backyard, the more rugged I can claim to be. I live with wolves, or I vacation with wolves, hence I am close kin to Kevin Costner. It rubs off. That's how consumerism works in an age when we're light-years past food and clothes and shelter.

In Abruzzo, Italy, where an active recovery plan brought the wolf back from the edge of extirpation (in a park one-sixtieth the size of the Adirondacks), 2,500 tourists a year used to visit Civitella Alfredena; now they've opened a wolf museum and 100,000 come annually. Wolf tourism is one of the big advantages cited by the environmentalists who want to reintroduce the animals, but there's something a bit creepy about it. The biologists at Algonquin say it doesn't damage the wolves to have organized howling parties out night after night in the summer, but who knows? It might damage the howlers—it's only an inch away from paying some sleazeball aquarium owner for the chance to swim with their dolphins and caress their blowholes. It's consuming, literally, the time and the attention and the wildness of another creature for some little jolt in return. For a story to tell when you get back from your trip. *It's about you.* The wolf becomes one more thing to experience, to own in some way or another. We live in a culture capable of taking a Civil War battlefield and transforming it into a theme park. We live in a culture fully capable of taking a wolf howl and turning into a commodity.

But maybe we still live in a culture capable of more than that.

Most Civil War battlefields haven't been handed over to the theme-park developers, thank heaven. Many of them, in the custody of the Park

Service, lie in that fast-suburbanizing corner of Maryland and Virginia and Pennsylvania, curious eddies of open space in a tide of pavement that has swept by on all sides. They have very little to do with that culture—the battlefields speak not to convenience or comfort, but to honor, obedience, glory, to that whole compound of older martial virtues now mostly mysterious to us. When we stroll Bloody Angle or the Sunken Road, when we climb Seminary Ridge, we visit powerfully different ideas of what we might be. Men gave their lives here for the Union? For Virginia? It's not that these were necessarily wise ideas—it may be far better to think of Virginia as an interchangeable place with San Jose or Tampa, a place to live until it's time to move. Glory and honor and obedience have caused their share of pain over the years. But the mere existence of these spots helps keep these alternate conceptions of human life alive and real.

I grew up in Lexington, Massachusetts, and I spent my summers in a tricorne hat, giving tours of the Battle Green. Some part of my political restlessness comes from endlessly reciting the story of the eight men killed on that common, men who inhabited a moral universe profoundly different from the moral universe of the twentieth-century Boston suburb that was otherwise shaping my soul. I remember the night in 1973 when I was twelve and I saw my father and five hundred other good suburbanites arrested because they refused to leave that Green during a protest of the war in Vietnam. They might have behaved the same way on some anonymous patch of ground, but they might not have, too; for an evening, two centuries of time collapsed, and Captain Parker's admonition to his troops ("If they mean to have a war let it begin here") nerved up another militia. We journey to Rome, to Mecca, to Benares, to Antietam, to Birmingham, and to the shiny black wall of the Vietnam Memorial with the sense that something there might change us. A pilgrimage, by definition.

So if we should be wary of how our endless neediness might shape our encounter with the wolf, we should be equally open to the idea that the wolf might be one force able to help us to reshape our endless neediness. The encounter with history at a battlefield or a shrine might be like the encounter with real wildness in a forest or a mountain or a seacoast; each might hold the power to break us out of the enchantment under which we now labor, the enchantment of things, an enchantment that is quite literally wrecking the world.

In fact, I would go further than that. Contact with the natural world is one of the two forces potentially powerful enough to break through the

endless jamming static of our culture and open us to other, wider possibilities. (The other force is contact with the non-possessive forms of love, be
they found at Fredericksburg or in Mother Teresa's hospice). I have seen
this effect in enough people, myself sometimes included, to be confident of
its truth. Show me an environmentalist who did not start with an encounter with the more-than-human world, and I will show you an exception to the rule. And try to find someone who has had such an encounter
without it changing her politics, her priorities, her very sense of self.

This encounter need not be wild in the traditional sense—it need not
involve Gore-Tex. Consider E. O. Wilson and his ants, or Rachel Carson
wandering her short stretch of the Maine coast. Consider the birdwatchers
of Central Park, necks eternally stiff from their involvement in the modest
life above them. For those of us with thicker shells, however, some starker
wildness, some drama, helps kick the message home. John Muir, on his
Thousand Mile Walk to the Gulf, found himself fixated on the notion of
alligators; later, in his exuberant first summer in the Sierra, a season that
defined forever our grammar of wildness, it was earthquakes and wind-
storms as well as sunny afternoons and butterflies that helped him understand that "we are now in the mountains and they are now in us, making
every nerve quiet, filling every pore and cell of us."

I've had some of the same sense of dissolving into the world on days
when I've stood staring at grizzly tracks in Alaskan mud. But I felt it much
more one day in the woods behind my house. I was wandering along,
happy for the exercise but lost as usual in some plan, lost as usual in my
own grandness. Suddenly, the fiercest pain I've ever felt boiled up my torso
toward my face. I whirled, staggered downslope, cracked into a tree, fell;
by now I could see the yellowjackets coating my t-shirt. I ran, flicked, ran;
eventually they were gone. But I could tell I was reacting to their venom.
Hives popped out across my upper body. All I could think of for a minute
were all the risks I'd heard dismissed by comparing them with bee stings:
"you're less likely to die by tornado/plane crash/shark attack." But as
I jogged back through the woods the few trailless miles to my house, I
found those thoughts replaced by another almost overpowering urge. The
urge to pray, and not a prayer of supplication but a prayer of thanksgiving.
The trees had never looked more treelike, the rocky ridges more solid and
rich, the world more real. The drama had somehow taken me out of myself, and though the sensation faded as the weeks passed, it has never disappeared altogether.

As I tried to describe the experience to others, I would say it was the first time I had felt a part of the food chain. But that was glib. In some way it was the first time I'd felt a part of any chain other than the human one. The first time I glimpsed the sheer overpowering realness of the world around me, the first time I'd realized fully that it was something more than a stage set for my life.

A healthy wolf has never attacked and killed a person in this country—yellowjackets are several orders of magnitude more dangerous, and let's not even discuss the Durangos and the Blazers and the Pathfinders. In a way, though, it's almost too bad that news is spreading of their benign image, almost too bad that people are starting to think of them as cuddly. Too bad because it's not true (on the basis of my hotel ballroom encounter, I will say I have rarely seen a being more aloof), and too bad because some of their power to shake us from our enchantment comes from their dramatic image.

Beavers just happened to be extirpated from the East; it was their poor luck that fashion found a use for the skin with which evolution had equipped them. But wolves were extirpated systematically, based on our understanding of their wolf nature. No creature save perhaps the snake has been more unremittingly vilified—the little pigs cowering in their shoddy homes, Little Red Riding Hood deceived by the granny-killer, werewolves snarling from page and screen, the very language making them the image of poverty ("the wolf at the door") and of the predatory male.

As much as anything else, that's the reason we need to invite them back in. Not to make it up to them—wolves roam outside the culture of victimhood—but because it is hard for us to do so. It would be a meaningful gesture.

You must remember that the Adirondacks, the Whites, the Greens, the North Woods of Maine are four or five or six hours drive from Madison Avenue, from the epicenter of the idea that everything should be easy. They are four or five or six hours north of the first suburbs on the planet, those ever-expanding monuments to the goddess convenience. They are due north of the megalopolis, of the notion that humans might survive in an entirely human corridor. They are, in other words, within range of the problem that needs solving.

I've been circling the exact nature of that problem, for it is daunting to search out the central point of a culture, but to state it as clearly as I can:

we believe too firmly that we, each of us as individuals, are the most important things on Earth. This belief is so strong in us that we begin to imagine
that things we have not created are not quite real. Thus, we can fret with
real worry about dangers to the economy (the chance, say that the Y2K
computer bug might damage it) but we cannot muster the same concern
for the natural world, which seems to us more abstract. Certainly we cannot muster enough concern to actually change the ways in which we live so
as to protect that natural world. Only the largest jolts break through that
shell, allow us to feel small in any way. Look at the few books about the
natural world popular at the moment; invariably they concern climbs up
Mt. Everest (and only deadly climbs at that), or storms producing hundred-
foot boat-eating waves; consider the vogue for adventure travel. How much
has changed even in the course of a hundred years, since the time that gentle John Burroughs, with his stories about woodchucks and chickadees,
reigned as the most popular writer of any sort in the nation.

Which in turn reminds us of how new this particular problem is, at least
in certain ways. Humans have always tended toward selfishness, always
thought too much of themselves—that's the message from the Buddha,
from the Christ. But for most of human history, we had other problems
too. No one living in the Adirondacks a century ago, or in any part of
America fifty years before that, was in any danger of forgetting that the
world was real, any more than someone inhabiting the delta of Bangladesh
at the moment is likely to forget. Our forebears faced the very real problem
of individual survival—of getting food on the table—and so, even with
the benefit of hindsight, it seems wrong to condemn them for chasing the
wolf away. Theirs was still the age-old problem that humans are slow and
hairless and dull-sensed.

But that's not *our* problem. In this country, at this moment, our problem is somehow reining ourselves in, figuring out that each one of us does
not reside at the exact center of the universe. Some years ago I wrote an
odd book that required me to watch everything that came across the nation's biggest cable TV system in the course of a single day: 2,400 hours of
videotape. If you condensed all those quiz shows and sitcoms and com-
mercials and newscasts down to a single idea, it was that each viewer is the
most important entity imaginable. That is an idea that would have at least
been questioned in most cultures in most places in most periods (cultures
that paid more attention to gods or to tribes), but it is the very bulwark of
the consumerist society we've built. We marinate in that idea—the idea

that there should be no limits placed on us. And we have the technology
to annihilate those limits at least temporarily. That is a dangerous combination. Many of the dangers of our moment—environmental, societal,
psychological—flow from that sense that we're all that counts.

So it comes as a piece of great irony—and maybe great luck—that the
part of the planet most committed to the idea that ease and comfort and
convenience are all that matters is also the place where nature has returned
most visibly in the last hundred years. The eastern United States, birth-
place of suburb and highway and car culture, co-creator with Hollywood
of the mirror we gaze into so fixedly, is also the only large region of the
planet that has gone from brown to green in the twentieth century. Once
topsoil was discovered in Midwest, the attraction of farming in New England, not to mention its economic viability, all but disappeared. Farms were
abandoned and trees returned. You could no more have reintroduced the

wolf to Civil War–era New Hampshire than to the moon; but now its land-scape is so much less manipulated, its forest so much denser, that the real possibility exists. We live in a biologically fortunate region, with enough humidity to erase some of the excesses of our past; if this isn't primeval for-est, it's pretty prime nonetheless.

Here, then, we can contemplate reversing time a little, having a second chance, this time not driven by sheer survival needs. We can contemplate putting ourselves in the background enough to allow, and even encourage, the return of a difficult example of the more-than-human, a creature that won't attack and kill us but might possibly take a lamb here or there, might even take a pet dog or cat, might force certain restrictions about land use or road closure on us, might provide some competition for our deer-hunt. And a creature that, even more, might excite us in certain new ways, open us more fully to the beauty and intrinsic meaning of the more-than-human world. If the wolf was not a difficult animal, then we could simply consume it, as we too often consume, say, loons. If the wolf was not in many ways a thrilling animal, then it would not offer as much promise for opening us to the real. But the wolf is both; when it howls, the hair stands up on the back of your neck for a variety of reasons.

To most effectively nudge our culture off balance, we should reintro-duce the timber wolf to White Plains, not the White Mountains; to Scars-dale and not Saranac. But that's asking rather a lot of a shy animal. In any event, it's not as if we who live in the mountains have escaped the general drift of our times. Adirondackers often live in the world that pulses through the satellite dish more fully than the world that surrounds them; most kids make it through high school without spending much time in the outdoors. Our life is increasingly unreal too. And if we would pay most of the price for this experiment—our pet dogs at risk—we would reap most of the benefit too. We'd hear the wolf howls year-round, not just in August.

By now the conceit of this essay should be apparent. It's not wolves that stand in need of restoration; wolves, though chased to the fringes of the continent, have managed to retain their essence. Even tied on leashes in the windowless basement ballroom of a slightly shabby hotel they were what they were.

People, on the other hand, especially the subspecies that inhabits the Western world, and especially the troop of them that has grown up on the eastern edge of North America, find themselves badly out of balance, operating under the influence of a spell. An intoxication, with ourselves. It's an intoxication that will likely grow deeper as the next few decades progress. Our unthinking consumption of resources has now reached the point where we are changing the global temperature, and with it the speed of the wind, the size of storms, the pace of the seasons. When we see the formerly natural world, we see ourselves. Our morally casual manipulation of genes may mean that within a very short time most of the organisms that we see around us will reflect our desires—will reflect us. We may be entering a period of overwhelming and unavoidable loss, of species extinction so profound that it may leave us with a planet almost empty of other meanings. The enchantment that prevents us from recognizing our peril as a peril, spiritual as well as pragmatic, and from making the changes necessary to correct it—that enchantment is our problem.

Happily, any child's bookshelf provides a series of recipes for dealing with dangerous enchantments. It's not the kisses of princes that will wake us up in time; our culture is one long wet kiss. But maybe the howl of a wolf might help, the howl of a wolf echoing out over the hills of the East.

An Ecologist's Perspective on Wolf Recovery in the Northeastern United States

John B. Theberge

THE DOMINANT ENVIRONMENTAL FORCE ACTING UPON wolves almost everywhere in the world today is—us. Guns, snares, bulldozers—we overwhelm wolf-ecosystem relations, abrogate the influence of their finely tuned social behavior, and insert ourselves into their predator-prey dynamics. Whether it is a wild population in the arctic or Algonquin, a recovering one in Minnesota or Wisconsin, a reintroduced one in North Carolina or Wyoming, humans are the predominant selective force causing wolf deaths. So much for real wilderness!

Wolf recovery in the northeastern United States is more an issue of what sort of wolf-like animal to get back, living what sort of life, than an effort to restore what is gone, because, ecologically, "we can't go home again." Biologically, genetically, wolf recovery is more than presence or absence— especially in the eastern forests.

We study wolves, and pick up their carcasses, in an environment that is geographically and ecologically closest to that of the northeastern United

States. In Ontario's Algonquin Provincial Park, wolves live where sugar maples, American beech, and yellow birch mantle the lake-studded hills, and balsam fir and spruce darken the lowlands, not unlike north-central New York State. The wolves of Algonquin know white pines and beaked hazel, and hear red-eyed vireos sing from the forest canopy all day and hermit thrushes usher in the night—just as they would in the Green Mountains, central Maine, or northern New Hampshire. Algonquin wolves hunt white-tailed deer, beaver, and moose, as they would there. And, most importantly, Algonquin wolves live in close association with humans. In eastern North America, wolves have no choice.

Jocko 10 died in late May 1999. A big, tawny-colored wolf with red-brown highlights across his shoulders, dark muzzle, and amber eyes, he had been the alpha male of the Jocko Lake pack. For four years we had radio-tracked him in Algonquin Park, through the forested hills he knew as home. We had radio-collared his father, mother, two of his sons, and two daughters. We knew his various den sites—one year a cave in a rock wall, another year a hole dug among the roots of a big white pine. On summer nights with the stillness otherwise broken only by the rasping of long-horned grasshoppers, we recorded his sonorous howls and those of his pack. The wolves often broke into chorus when pack members left or returned to their beaver-meadow rendezvous sites; we knew from the beeps on our receiver.

Jocko 10 went missing in January 1999. We kept his frequency programmed into our receivers along with those of the other collared wolves, searching from the air and ground. Then, on a flight farther southwest than normal, his signal came in—on mortality mode. His collar was just outside the little town of Madawaska.

Mary, my co-researcher and wife, and I took cross bearings on the signal from a highway and discovered that it came from a dilapidated shed behind a house. Nobody was home, but we returned that evening and found a teenager working there. We asked about the collar, and after visible initial shock, he feigned ignorance. He said his father was just down the road and would be back soon, so we waited in our truck. Almost immediately, the signal went off mortality mode—the boy had moved the collar. Then he came out, climbed into a pickup, and roared away, the signal fading with him. Minutes later he returned, went back in the shed, and came to our truck window with the collar in hand. He explained that his father had found it lying on an abandoned railroad and had put it in the shed. He professed knowing nothing about it until then.

Picking up radio collars of wolves killed by humans is an integral part of wolf research, anywhere. We have been doing it for twelve years. Some people who kill wolves think we will never find a radio collar in their shed or house. Other people smash them. Some deny having killed the wolf. Others run to the local newspaper to brag about their exploits. There is a whole sociology of wolf killing waiting for someone to study it: age, occupation, education . . . all those things. Such a study could be carried out in Alaska, Yukon, Wyoming, northern Ontario—it probably doesn't matter where. The results would be similar. A lot of people have fixed attitudes about wolves.

What we have found out in Algonquin serves as a rough guide to answer questions about wolf recovery in the northeastern United States. How would the ecosystems be different if wolves were there again? How might wolves influence their prey, and vice versa? How would wolves space themselves, and what densities would they reach? What low numbers might still allow them to persist? Do they really stand a chance in the face of the inevitable boardroom bickering and backcountry politics?

The dictates of science have invariably shaped our study and the way we think about wolves. A scientist's job is to gather evidence, sift through facts, and derive conclusions. Science is only a structured way of thinking, but one that provides a confidence-building foundation for understanding nature, one that separates the anecdotal from the quantifiably testable. It increases the probability that opinion approaches reality. So we have played by the rules of reducing events to empirical data, supplementing field notebooks with data sheets, setting up hypotheses for testing, and building up statistically significant sample sizes.

There is a certain rigidity to reducing activities and experiences out in the woods to "data" in that way, and some say it causes scientists to lose perspective, not recognize connections, regard the animals they study as mere numbers. That can happen, an occupational hazard, but to avoid such "desensitization," you just have to be aware of the danger, like avoiding an automobile accident. If the student of nature maintains an open mind, approaching nature from a scientific perspective deepens, rather than supplants, a sense of wonder. As Rachel Carson expressed in *A Sense of Wonder*, "Once the emotions have been aroused—a sense of the beautiful, the excitement of the new and the unknown, a feeling of sympathy, pity, admiration, love—then we wish for knowledge about the object of our emotional response." Wonder breeds respect; respect makes you care. We were attached to Jocko 10 who, like so many wolves, met a senseless death by human hands.

Science offers more than some degree of certainty about nature. From its various subdisciplines it supplies time and space context. Every species, population, and ecosystem is surrounded by a geography and trails a history. How would wolves fit in if they came to thread the forests of the Northeast once again?

WOLF TIMELINE

Long before humans appeared on Earth, a variety of wolf-like creatures came into being. A wolf-like design has been hugely successful: 50- to 100-pound animals built for speed, teeth adapted for piercing and shearing, jaws strong enough to crush bone. Through the last 40 million years, most of the Age of Mammals, few ecosystems on Earth have been without one or another of them. Parading across time, shaped by landscapes and prey species long gone, first came the dog-faced cynodonts at the early edge of mammalian diversity in the late Mesozoic Era, then the Order Carnivora in the early Cenozoic. The Carnivora radiated into the Family Canidae 45 million years ago with an array of increasingly more specialized dog-like forms, including Oligocene species such as Cynodictus and Amphicyon, then Miocene species living 20 million years ago such as Cynodesmus and Daphoenodon, the latter with remarkably wolf-like skulls. These early dog-like species in turn gave rise 12 million years ago to the hyena dogs, with broad skulls and powerful jaws, which dominated for nine million years right up to the early Pleistocene. As the hyena dogs declined to extinction, the stage was set for the radiation of Genus *Canis*: *Canis etruscus* in early Pleistocene Europe; possible coyote progenitor *Canis lepophagus* in North America; *Canis edwardii*, the first unquestionable wolf in North America, some 2.5 million years ago; its possible descendent the larger *Canis armbrusteri*, which ranged across much of North America more than a million years ago, but apparently was restricted later to the East before bowing out about 500,000 years ago. Before it left, it may have given rise to the even larger dire wolf, *Canis dirus*, or that wolf may have originated in South America and migrated north. And about one million years ago give or take a few hundred thousand years—it simply is not clear from this *Canis* taxonomic cauldron—somewhere out of all these *Canis* forms arose the wolf as we know it. *Canis lupus* and its possible immediate forerunner, *Canis rufus*, along with the coyote, *Canis latrans*, are the surviving wild North American lineages, honed by changing environments, and the panoply of species that served as prey, and time.

One could argue that a wolf-like predator, irrespective of species, is a basic part of any ecosystem with large, herbivorous mammals. Only in South America did the cats more completely fill the canid niche. Evolutionary history is full of ecological replacement. Species are not immutable; they come and go. There will be more. The canid evolutionary forge designed to make wolf-like animals has remained intact over the past tens of millions of years, through ice ages and climate warmings and the comings and goings of various species. That forge is still unbroken and will be as long as 100- to 1,000-pound herbivores still live on Earth and wolf-like genes are around to remold and reshape.

Someday, wolves or a wolf-like descendant invariably will repopulate what is now the Northeast, regardless of us. The second 5 billion years of planet Earth's history is still waiting to be written—plenty of time, assuming we are not here to impose our repressive will on evolution too much longer.

Our knowledge of the recent history of genus *Canis*, like its distant origins, rests on a weak fossil record. Science deals not just with proven facts but with speculation based upon available evidence. Most of the evidence comes from skull measurements and genetic testing. The most widely accepted history proposed for wolves comes from years of painstaking work by morphologist Ron Nowak, recently retired from the United States Fish and Wildlife Service.

First of the currently living wolves in North America appears to have been the red wolf, *Canis rufus*. Some of these wolves may have migrated across the Bering land bridge to Eurasia, because wolves show up in the fossil record there. Alternatively, wolves arose there, but in either case New World and Old World wolves must have been separate, or largely so, for a long time. Recent genetic evidence from Brad White's laboratory at McMaster University in Hamilton, Ontario, indicates a separation of one or more million years, certainly enough time, if sufficiently isolated, for at least subtle differences, or maybe even genetically distinctive species to arise.

Possibly about 300,000 years ago, what we identify now as the gray wolf, *Canis lupus*, entered North America and populated most of it. At least three subspecies spawned off these immigrants: *occidentalis* that was likely isolated by late Pleistocene glaciation in the Northwest; *arctos*, isolated in a glacial refugium in the Arctic Northeast; and *baileyi*, isolated by desert conditions in the South.

Besides these three subspecies, Nowak classified two others: *nubilis*, likely the species, or descendant of the species, that came from Eurasia; and *lycaon*. The origins of *lycaon* have remained a mystery because no obvious geographical barrier to foster its formation has been found, as there has for the other subspecies. Brad White, Paul Wilson, and their colleagues have provided a conceivable answer. Genetically, the red wolf, *Canis rufus*, and the *lycaon* subspecies of the gray wolf are similar enough to be considered the same species. Brad White does not consider *lycaon* to be a gray wolf at all. If true, *lycaon* likely was here from the outset as the North American–evolved wolf. Accordingly, White, Wilson, and colleagues have proposed the reclassification of both the red wolf and *Canis lupus lycaon* into the same species, *Canis lycaon*. Historical precedent supports that name: *Canis lycaon* was the first name given North American wolves in 1775 by European naturalist Johann Schreber.

This discovery is relevant to wolf recovery in the Northeast. Based upon White's genetic results and Nowak's skull measurements, the red wolf and *lycaon* historically had contiguous and overlapping ranges from Florida to just north of Algonquin Park. Today, New England and New York State are geographically sandwiched between the current red wolf range to the south and the *lycaon* range to the north. The red wolf, extirpated in the late 1960s from the wild, was reintroduced in 1988 into a wildlife refuge in northern North Carolina. The *lycaon* wolf lives in central Ontario, including Algonquin Park, and southwestern Quebec. *Lycaon*, then (or the red wolf/Algonquin wolf/eastern forest wolf/North American–evolved wolf —all the same species), is the logical candidate for restoration in the Northeast. Currently, we have no evidence that the gray wolf ever lived in the Northeast. Supporting this interpretation, a skull collected in Maine in approximately 1867 has been identified on the basis of both measurements and genetics as *lycaon*, not *lupus*.

A contending theory, however, indirectly favors the gray wolf for restoration in the Northeast because there is no truly different wolf species. The theory was put forth previous to White and colleagues' recent genetic work, by Robert Wayne, N. Lehman, M. Roy, and colleagues in California.

They proposed that both the red wolf and *Canis lupus lycaon* are not true species but rather are gray wolves that have hybridized with coyotes. This conclusion is based upon the presence of what they identified as coyote-like mitochondrial DNA in both red wolves of the southeastern United States and in *Canis lupus lycaon* wolves from southern Ontario and southern Quebec. Mt-DNA is found in the small, granular-like mitochondria of cells situated outside the nucleus, site of much cellular metabolism. Wayne and colleagues based their interpretation of hybridization on a greater molecular *similarity* to existing coyote, not its exact replication. Mt-DNA is cloned and passed down from the female—rarely is there any male contribution. According to this hypothesis, female coyotes have bred sufficiently with male gray wolves to result in a hybrid position for both red and *lycaon* wolves.

Both genetic camps agree that typical gray wolf mt-DNA is absent in both red wolves and *lycaon* wolves. The White camp, however, interprets about 75 percent of what the Wayne camp called coyote mt-DNA as species-defining *lycaon*/red wolf–specific mt-DNA, because it does not exist in any living coyote populations. In addition, the White camp has gone on to examine microsatellite DNA residing within the nucleus of cells. They graph out very similar and overlapping bell-shaped curves for the red wolf and *lycaon* wolf based upon an individual animal index, which strongly supports their conclusions that the two are the same species. Furthermore, these curves are distinctive from those for either coyote or gray wolf.

If *lycaon* and *lupus* are different wolf species, the implications for conservation go beyond the Northeast. This interpretation also influences the red wolf recovery program in North Carolina, until now thought to be the only living wild population. As well, it complicates a pending decision in Minnesota to declassify the wolf as an endangered species. There, the two species may actually be living in close association. Wayne and colleagues classified all Minnesota wolves as *lupus*, subspecies *nubilis*, with half of them hybridized with coyotes. But bothersome about this interpretation is that nowhere else in North America except Minnesota does *lupus*, subspecies *nubilis*, hybridize with coyotes, even where they live in the same habitats throughout much of wolf range in the West. However, if what Wayne considers hybrids are really *lycaon* wolves, as White and colleagues suspect, then the hybridization phenomenon becomes more understandable. It would then only be documented between coyotes and eastern forest wolves, specifically red and *lycaon*. Before declassification, some additional genetic analysis needs to be undertaken in Minnesota, Wisconsin, and Michigan to clarify the range and extent of the two wolf species instead of treating them as one.

Scientific acceptance of new theories quite often is a process, not an event. Other researchers usually want to run tests and try to replicate results. Then, over time, if the new seems more convincing, it is referenced more often and soon becomes "common knowledge." While we may never completely parse out the wolf's evolutionary history, we can hope, for the sake of conservation, at least to discern what species and subspecies are alive today.

WOLF SPACE DIMENSIONS

The wolf is a generalist, once enjoying the largest range of any land mammal, not only in North America but in the world. Species, however, are not simply distributed at random across continents. The newly emerging field of "macroecology" emphasizes pattern at a large scale. In North America, for example, the density of species living in the South is greater than in the North, as it is at any given latitude in the West versus the East.

Various theories have been proposed to explain this pattern. They range from the scouring effects of glaciation that periodically reduced species in the North, to habitat variation driving higher species diversity in the mountainous West, to the influence of predators in tropical areas reducing competition between prey species that would otherwise exclude one another.

Regardless of theories explaining why, the existence of this pattern illustrates that the number of species living in any area is not simply the product of the local environment, as tempting as it is to consider only that. Explanations for levels of biodiversity must be sought at different scales: local, regional, even continental.

James Brown, is his book *Macroecology*, compared the range dimensions of North American mammals by graphing their east-west against their north-south extents. He found that a disproportionate number of species with large ranges extend farthest in an east-west direction. His speculative explanation was that species with large ranges tend to be limited primarily by the large-scale patterns of climate and vegetation that run in broad east-west belts.

The wolf, however, breaks the rules, spanning both the full east-west and north-south dimensions of the continent. Its biological design has kept it from being cornered by geography just as it has from being cornered by time. Globally, it is, or has been, found almost everywhere once connected by a land bridge between North America and Eurasia except for the hot tropics and hot deserts while its presence in places such as Saudi Arabia pushes the latter. It takes a barren environment to exclude the wolf. It is the ecological generalist among generalists, seemingly insensitive to climate, vegetation, or other local or regional conditions. It asks little— only freedom from persecution by humans and the presence of any species of large herbivorous mammal to eat. It trots across the tundras of the high arctic, threads the mountain passes in western Canada, and lives in the temperate rainforests of Vancouver Island. It has holed up and is surviving in India, China, Israel. . . . It has repopulated portions of Rumania, Croatia, Spain, Portugal, Germany, and France. More instructive in understanding the wolf's distribution is to look where it is absent: anywhere urbanized or where it has been snared, shot, trapped, or poisoned out of existence. That, unfortunately, encompasses most of the United States and southern Canada.

With adequate protection, wolves could conceivably come back to almost all but the heavily urbanized areas of their former range. And, in all the places where they have returned—Europe, North Carolina, Idaho, Wyoming, Montana, Michigan, and Wisconsin—they have received belated legal protection. One searches in vain for examples of wolf expansion into areas where they have not.

Natural re-colonization is a consequence of "metapopulation," one of the important space-related ideas in ecology. The concept emerged from a famous theory on island biogeography first put forward in the 1960s and 1970s and mulled over and tested repeatedly since then. Originally developed from studies of islands, the theory expanded into general principles applicable anywhere. Most continental environments consist of various habitat patches that isolate especially the less mobile species. The theory emphasizes the importance of periodic re-colonization—immigration from somewhere else—for the persistence of most populations, which is most probable if supply populations lie nearby. Without periodic immigration, ecosystems experience "decay"—an incremental loss of species. Small, isolated populations slip into the extinction vortex, like water running down a drain.

"Small" for a wolf population lies in the order of two hundred animals for long-term genetic viability, according to a number of calculations. Called "minimum viable population," such a calculation is based on the well-accepted principle that no population should experience greater than one percent inbreeding per generation. As well, when a population is small, many catastrophic climatic events may threaten it: ice storms, wind, deep snows, drought. Local over-harvesting can drop numbers too low for a population to recover, as can an outbreak of disease, or a shift in predator abundance, or fire. Small populations have limited toolkits of genetic variability to offset environmental shifts due to climate, logging, or landscape change. Conditions like these may draw down small, semi-isolated populations all the time, but we don't notice because dispersing members from nearby populations immigrate in and tend to stabilize numbers.

Dispersal is one of the least understood aspects of ecology, not through lack of theory or study, but because of technical difficulties. For years, population ecologists have tended to pick study sites for almost any species on the basis of habitat homogeneity. In that way, changes in numbers can be studied as a function of only birth and death rates, while immigration and emigration are considered equal to each other and thus ignored. But they should not be ignored. Increasingly, they appear to play a fundamental role in the persistence of species almost everywhere, especially in the face of the types of environmental change so prevalent today.

The "metapopulation" way of perceiving populations is one of swirling connectivity across a species' range—local populations occasionally becoming reduced or even going temporarily extinct, infilling from elsewhere, ebbing and flowing along habitat corridors and patches. Unevenness in habitat quality dictates a lack of uniform population density and creates semi-isolated population "demes" where groups of individuals tend to breed more within than without. Abundances shift due to exigencies of local environments. If fragmented landscapes inhibit metapopulation flow, species suffer.

Even the wolf is put in jeopardy where populations are small and isolated, with no, or limited, metapopulation flow. Wolves introduced in 1960 on Coronation Island in southeast Alaska went extinct. Wolves on Isle Royale, famous because of more than thirty years of research, exhibit a 50 percent loss in genetic heterogeneity that could threaten their long-term survival.

The architects of wolf recovery programs in the United States have recognized the importance of metapopulation by considering a population established only if three subpopulations live in close enough proximity for periodic exchange of members. Success seems imminent for the gray wolf in Idaho, Wyoming, and Montana; it has been more elusive for the red wolf where the reintroduction in North Carolina worked but the one in Tennessee failed. Even more difficult has been establishing just one population of the Mexican subspecies of gray wolf, *baileyi*, in New Mexico and Arizona.

A single, isolated wolf population established in the Northeast would therefore be in jeopardy. Any possible link to the red wolves of North Carolina would be unlikely because of heavy urbanization. Equally unlikely is any link along the Frontenac Axis between Algonquin Park and the Adirondacks, despite a possible corridor south from the St. Lawrence River identified by Stephen Trombulak and Jeff Lane of Vermont, and suitable habitat running north to Algonquin Park. Considerable human development breaks this corridor on the Ontario side, and we now know that the axis lacks a wolf population all the way up to Algonquin Park. A possible connection exists between Maine and southern Quebec with corridors both east and west of Quebec City, identified by Daniel Harrison and Theodore Chapin, but these connections provide for little if any wolf movement now.

The wolf is a long-distance disperser capable of traveling hundreds of kilometers. It can exhibit metapopulation flow as it has in the upper Great Lakes states, navigating a wide variety of habitats. That ability enhances the possibility for natural recovery in the Northeast. The wolf cannot, however, negotiate urban sprawl or cross picket lines of hate.

THE WOLF AND ECOSYSTEM STRUCTURE

The wolf comes with various ecological labels, each representing an idea about its place in ecosystems or the impacts it levies. Two of the most common labels are "summit predator" and "umbrella species." They relate to ecosystem integrity, itself a buzzword in environmental management but more than that, an important goal. Have the forests of the Northeast lost ecosystem integrity because the wolf is no longer there?

The most obvious structure of ecosystems is their organization into trophic levels—primary producers (plants), primary consumers (herbivores), secondary consumers (first-level carnivores), tertiary consumers (second-level carnivores)—with a web of interactions among species living

at all levels. Rarely do temperate ecosystems such as northeastern forests have any strong representation beyond secondary consumers, at least among terrestrial vertebrates, except for raptors that eat insect-eating passerine birds. It is in this construction that the wolf is considered a "summit predator."

The most notable feature of this trophic structure is its energetic inefficiency. A mere 1 percent of the energy in any one level carries over as net energy in the next level ("net" is the energy captured minus metabolic and digestive losses). Life on Earth hangs by a thin energetics thread. Due to energy losses, there can never be nearly as many wolves as prey. In fact, predator/prey ratios in ecosystems with wolves, as with other large carnivores, are shockingly low. The biomass (weight) of predators generally represents less than 0.5 percent of the biomass of their prey.

This low ratio, coupled with a normally sparse density of prey—one moose per square kilometer is a dense population—means that wolf densities can never be very high. One wolf per 25 square kilometers was achieved in the 1960s in Algonquin Park. That was the density for many years on Isle Royale, bounding to double that for a very short time before a crash. Montane and low-arctic densities of wolves are in the order of one per 250 square kilometers. No place will ever get overrun with wolves, contrary to popular misconception.

Often linked with the wolf's position as a summit predator is the more philosophical idea that a land that can grow a wolf is a healthy land, one with all the ecosystem parts in place: prey species, and the plants that the prey species browse or graze upon, and the soil nutrients that support the plants. That perception of ecosystem health may be valid but does not mean that all the interlocking pieces of the ecosystem are natural. Chickens and goats are primary food for wolves in some European countries. Wolves scavenge in garbage dumps. Where not persecuted, wolves can live in greatly altered landscapes, even unwholesome ones.

So, rather than being diagnostic of the health of an ecological community, the wolf is just another member, with not too demanding requirements—except freedom from human persecution. Wolves, alone, in the Northeast will not make the land intrinsically healthier, just perceptually so. There still will be a need to prevent air pollution, and practice ecologically based land management, and legislate against toxins entering the waterways—still a need for vigilance.

The wolf is referred to as an "umbrella species" in the field of conservation biology, where scientists search to understand the management significance of biological traits. A species-by-species approach to conservation often is considered unrealistic because there are so many species. An alternative is to emphasize habitats, hoping that if representatives of all habitat types are protected, all species will be too.

But species may become rare or even disappear without any habitat alteration for many reasons, including excessive human persecution. Or habitats may become dysfunctional for certain species in subtle ways through upset species relationships, again often linked to negative human impacts.

Useful to conservation, then, is some combination of species and habitat approaches, and "umbrella species" provide that opportunity. The wolf is one because of its summit position in food chains, but, even more importantly, because of its wide space requirements. Take care of the habitat needs of a wolf population, goes the theory, and you automatically take care of the needs of the many smaller, less space-demanding creatures. Not always, of course, and so the idea has its critics, but often enough that the general principle is valid. How much land does that represent? About 45 square kilometres per wolf in Algonquin Park. Multiply that by the minimum size of a viable population, and that covers the needs of a lot of marten, beaver, red squirrels, and a few million chipmunks.

Taking a longer term, evolutionary perspective on the impact of the wolf on ecosystem structure inevitably focuses on animal behavior. All animals liable to be eaten either have evolved defense mechanisms or are no longer with us, an observation that says something about the importance of predation. The danger of getting eaten is not just a peripheral issue for most living things. Among the prey species of wolves, for example, beavers stay close to ponds; the risk of predation is likely the very selection pressure that favored pond building in the first place. Wolves and other predators of beavers, therefore, are important architects of biodiversity, because beaver ponds provide habitat for so many species. Moose tend to calve on islands, peninsulas, or hilltops due to the dangers of wolf predation.

Every potential prey species large enough for a wolf-sized animal to hunt either climbs trees, digs dens, has quills, tastes bad, runs fast, or builds ponds. The wolf cannot take all the credit for this important structuring of other species. the organisms that we see today have made it through a long evolutionary filter that involved other predators such as saber-toothed cats, dire wolves, and cave bears, as well as all the wolf forerunners.

Today we live in an impoverished world for large predatory species compared with the Pleistocene. The wolf alone, in most places where it lives, is charged with the responsibility of keeping up the selective pressures on potential prey—except for predatory humans. Where wolves are absent, should we concern ourselves about a lack of wolf-conveyed selection pressure? We have yet to expand conservation to encompass the qualitative characteristics of living things.

Yet another way to look at ecosystem structure is from the nesting of species in a hierarchy of size. All ecosystems harbor small-bodied herbivores or omnivores that live in small territories and eat small, high-energy items like seeds, buds, nuts, or insects, grading up to large-bodied herbivores and omnivores that roam over large areas and can eat coarser foods. All have attendant predators showing a similar range in body sizes. In that fundamental way, ecosystems are partitioned so that many species can live together with limited direct competition.

A complete ecosystem, one with all the expected species, exhibits a full range of body sizes among its mammals. Gaps represent species lost or vacant niches. Most commonly, gaps are found among the large-bodied species. These are the most extinction-prone. Uniformly across extinction events, whatever their magnitude, large-bodied species have suffered a greater proportion of extinctions than small-bodied ones. Reasons, described in many studies of this phenomenon, include later sexual maturity, smaller litter sizes, lower population densities, or in some cases more pronounced habitat specificity—in general a more conservative survival strategy than the "flood-the-world-with-young" strategy of smaller species.

Wolves employ a conservative reproductive strategy, with longer dependence of young and later first breeding than the herbivores they hunt. Coupled with low density, wolves are inherently at risk. Their broad habitat tolerances may buffer this vulnerability, but that trait does not help them offset the effects of heavy mortality. And, once gone, a vacant niche may not be guaranteed indefinitely for their return. Ecosystems adjust. They have in the Northeast, as I will describe later.

THE WOLF AND NICHE SPACE

The toolbox of nature comes complete with all the species necessary to assemble an ecosystem, those species with shapes and sizes, colors and characteristics to fit others around them; those just right for a particular climate, topography, and soil; those crafted by time to capture energy and suck up their share of available nutrients in various and different ways. Despite some conditions that occasionally permit species to occupy the same niche, the economy of nature rarely allows complete substitution, at least not concurrently. Species fit roles, not just places in their environment—they do certain things such as eat certain-sized food in certain places at certain times.

The wolf has been called a "functional ecological dominant" in an effort to describe its role in ecosystems. It holds that title because of its position as a summit predator and because it can levy a significant impact on the environment, more than most species. It can, at times, exert considerable control on the characteristics of ecosystems.

The extent to which the wolf is a functional ecological dominant depends most immediately upon its influence on prey numbers. Many studies have attempted to assess this influence, but with difficulty. The problem lies in what ecology books call "compensations" made by the prey population that tend to offset the impact of predation.

"Compensatory natality" is one of them, an increase in birth rate that offsets any increase in losses. The mechanism usually involves better nutrition for the remaining individuals, and, in some species, a resultant decrease in intraspecific competition. As with weeding a garden, removing some individuals makes room for others to grow.

Some prey species are more capable than others of exhibiting compensatory natality. With good nutrition, both moose and white-tailed deer produce twins, and females breed at an earlier age. Caribou, however, never produce twins. Beavers in some studies, although not all, have been shown to produce larger litters where trapped. In response to exploitation, muskrats can produce multiple litters in a summer.

Even more significant as a mechanism to offset losses is "compensatory mortality," the notion that if one thing doesn't get them, something else will. If predators are not present, more prey might starve, or become diseased, or fail to breed because of social dynamics. Particularly important is the question of whether young animals killed by predators would have died anyway; mortality among young, with their host of pitfalls, is normally high in most wildlife species.

Designing research to test the operation of compensatory mortality is difficult. The most common approach when studying small mammals or birds is removal experiments, with a companion "hands-off" control area for comparison. But when studying a wide-ranging summit predator like the wolf, the necessary condition of ecological equivalency of two very large areas is never met. The best that can be done is to draw inferences by comparing rates and causes of mortality over a series of years.

Some years ago, David Gauthier and I summarized the results of various wolf studies up to 1985. Without always addressing compensatory mechanisms directly, slightly more than half the studies presented evidence that could be interpreted as wolf limitation on the population size of one or more of its prey species. The other studies concluded that other environmental factors were more important. These results suggest that the impact of wolf predation on prey populations is inconsistent, but that seems logical. Conditions other than predation must be expected at times to limit ungulate populations, especially human killing, disease, and range quality as influenced by climate. On the range edges of a species, for example, habitat conditions that prevent further spread normally also limit numbers.

The wolf is a functional ecological dominant if the ecosystem operates predominantly from the top down, that is, if the wolf influences the numbers and distribution of herbivores, which in turn, influence the species composition of forest or grassland. Top-down control may be most pervasive under relatively stable environmental conditions. However, the wolf is not so dominant, becoming a "free-rider" that exerts little influence, if the system operates predominantly from the bottom up. This condition is more common with frequent vegetation disturbance such as forest fires or outbreaks of spruce budworm. Then, vegetation and climate primarily influence prey numbers and distribution, which, in turn, influence wolves.

Ecosystems are complex, multivariate, dynamic. Snow, rain, cold, heat, forest insects, animal diseases, logging, hunting, and trapping all intertwine to alter morality, natality, emigration, immigration. Perhaps the very top-down versus bottom-up question is too simplistic. In broad terms, though, whenever an ecosystem is subjected to heavy human pressures, regardless of the initial prevailing direction of the controlling influences, they get jumbled. Likely this jumbling of relationships, this destabilization, represents the most substantial environmental impact that humans exert on natural ecosystems. For example, human-caused mortality that draws down wolf numbers will influence predator/prey ratios and may tip the balance of control from top-down to bottom-up. Or, a significant drawdown of ungulates will cause the reverse, and may result in the release of seedlings of preferred browse species such as red maple that may have been suppressed by ungulates, and at the same time may induce wolves to prey more heavily on beaver, as has been shown repeatedly in wolf food-habit studies around the Great Lakes. Impacts on one component of a system invariably reverberate throughout.

Regardless of the predominant direction of control, wolves are not essential to the functioning of ecosystems. Ecosystems can do without them. As much energy and nutrients simply will not make it as far up the trophic chain before heading off to feed the detritivores. Populations of prey species will be limited by something else. Poplars farther from water will get cut down because beavers are not so pond-dependent. Moose won't bother to calve on islands or peninsulas or hilltops. No species is essential in ecosystems, none of the panoply of species that have paraded across the world stage over the ages—even us.

Without their summit predators, however, ecosystems are different. They still function, still provide home for many species. But if ungulates are more abundant because of an absence of wolves (not a sure thing), then forest species composition changes, such as a different ratio of pines to maples. Everything will adjust, from nesting birds, because of canopy structure, to soil nematodes, because of needle-influence versus leaf-influence on soil pH.

So the wolf, even as a functional ecological dominant, is only as essential to ecosystems as we hold the presence of all the native species to be. A complete complement of species is often used to define ecosystem integrity. But complete when? Ecosystems and species are ever-changing through their own processes of succession, erosion, deposition, climate variation, fire, disease. There is no answer from an ecological/evolutionary perspective. There is, though, in what people want, in perceptions and attitudes. The wolf was an integral part of the northeastern forests before it was extirpated by humans. It is still necessary for ecological integrity, regardless of the role it plays.

THE EASTERN FOREST WOLF

Jocko 10 was a typical Algonquin Park wolf, at home among ranks of rolling pine-poplar-red maple forests. In summers he often hunted near the shores of Jocko and Feely lakes where the loons make the hills ring; in winters he traced the white meanders of Carcajou Creek. In his territory, he knew every ridge, beaver dam, deer runway, rocky point, lakeshore, strand line, hemlock stand, and spruce thicket, because that knowledge meant survival. He had his own pack, his territory, various den sites among the pines where his mate whelped their pups, and a variety of mid- and late-summer rendezvous sites. He knew crackling cold winters and hot summers, knew where the moose calved and where to escape the heat and flies of a July day, knew when the deer migrated and where they went. He and his pack often migrated there, too.

He was born into the Jocko Lake pack in 1992 and never left. In the early autumn when he was one-and-a-half years old, his mother died of natural but unknown causes. That December, his father found a new mate and brought her back to his territory. When Jocko 10 was three years old, he competed for the position of alpha male and usurped his father's position—kicked him out. A few months later, his wandering father was shot outside the park.

That spring, his stepmother bore his pups, and again the following year. By then, 1997, he was the alpha male of a pack of six. In two successive years, one of his yearlings dispersed northwest to join the Travers pack, crossing one other pack's territory to get there. The Travers pack suffered a sad history of repeated mortality from humans and as a result was open to recruits.

In mid-winter 1997/98, his mate's radio collar went off the air. Our evidence suggested that she was a victim of human killing, too. That summer, Jocko 10 led his pack in a wandering pattern that centered on no den or rendezvous site. They spent much of their time on the lands of the Redpole Lake pack to the southwest. That pack had vanished the previous winter except for the alpha male, who occupied the very southern fringe of his former territory. The Jocko Lake pack may have kept him there, taking over his lands; often the Jocko Lake wolves were less than a kilometer from him. But then they would leave the Redpole territory and swing back to lace out some footprints on the logging roads they had used in previous years. They never stayed anywhere long.

Jocko 10 lost both a son and a daughter in the late summer and autumn of 1998, both to unknown natural causes. We radio-tracked them to their bodies but they were too decomposed to determine the causes of death. Jocko 10 did not have long to mourn. Only a few months later, during hunting season, he went missing. Packless, he must have wandered outside the park to his death. The teenager at the shed where we found his collar would not tell how or where he had been killed.

After we picked up his radio collar that May day, we left it on the floor by the front seat of our truck. There it stayed while we went about our work: radio-tracking other wolves and conducting our annual browse and pellet surveys. A few days later, we drove the logging roads into the Jocko Lake pack's former territory to see if other wolves had reoccupied the land; to search for footprints along the roads or crossing the mud of beaver dams; to listen for howls in the night.

We found no tracks, although we drove and walked many kilometers, and we heard no howls. At our feet lay the collar that had ranged all over that land around Jocko 10's neck. It seemed strangely as if we had only lent him the collar. Unwillingly, he had returned it. We didn't talk much. The excitement of wondering where he was and what he and the rest of the pack were doing was gone. When a hockey game is 6–0 for the other team, your interest lags. That is how we felt. For us, the land wasn't whole anymore. Sure, it would heal and other wolves would take over. But wolves, like people's dogs, like people, have personalities. The Jocko wolves no longer gave life to the landscape. We were glad to pull out when our work was done.

Eastern forest wolves, *Canis lycaon*, once graced wild landscapes from Florida to southern Ontario and Quebec, and from the Atlantic seaboard maybe as far west as a line between Oklahoma and Manitoba. They vanished from most of this range before the turn of the twentieth century, along with the eastern forest elk, eastern cougar, heath hen, and ivory-billed woodpecker. Only a sparse and anecdotal literature chronicles what the eastern forest wolves had once been like.

Unlike those other species or subspecies, however, they did not go extinct. Their adaptability has given them, and us, a second chance. As people increasingly recognize that there can be too much humanity, that the biosphere can absorb only so much plundering, that wildness is disappearing, discussions heat up about how the eastern forest wolf might be put back. The land still has room for it. American chestnuts may be gone, along with many of the giant old hardwoods and pines, but the other species that made up the historic forest associations are still there. Recent forest regeneration is healing the scars of an overexploitive past. At the turn of the twentieth century, two-thirds of New York State was nude. Now that figure has been halved. The same trend predominates throughout the northeastern United States.

In both Algonquin and the Northeast, not only similar species but also similar processes are at work. Deep snows trigger deer migrations to lowland conifers; beaver follow similar patterns—on land in the early spring before the aquatics develop in the ponds and in the fall when they stockpile winter food. Both regions include large blocks of public land with policies aimed at wildlife protection. There is still room for the wolf on the land, but is there room for the wolf in the human psyche?

Algonquin wolves have some answers. They are not necessarily encouraging, but they illustrate issues surrounding wolf persistence. Over the twelve years of our study, the wolf population on the eastern half of the park, where most of our work has been carried out, has declined by 35 percent, and since the early 1960s by more than half. Densities across approximately 3,000 square kilometers have fallen to 1.9 per 100 square kilometers, figures derived from repeated aerial observations and ground tracking each year. Wildness has been devalued accordingly.

Among the possible reasons for this drop in numbers is the status of their prey. Deer in Algonquin have declined to one-sixth their density in the 1960s, from close to six per square kilometer to less than one, even lower in the northwestern sector of the park. The deer declined through some combination of severe winters in the 1970s from which they never completely recovered, the removal of winter cover by logging, and most importantly in recent years, overhunting. Deer are killed when migrating or on their wintering grounds by native people inside the park and by white people outside. Recognizing the mistakenly high quota, the government recently decreased the allowable kill of 750 antlerless deer by 80 percent. We calculate that wolves, in comparison, take approximately half as many as that previous allowable kill—and some people resent even that. The wolf kill is low because they split their diet roughly equally between deer, moose, and beaver. While wolf predation may represent a significant mortality factor for deer, and possibly place a drag on their population recovery, it did not precipitate the decline.

Moose, in contrast, have increased since the 1960s by about threefold to a density of slightly less than one per 2 square kilometers. That is high by Ontario standards. In the mixed-wood ecosystems on the southern part of their range they find a combination of abundant food and good cover. Algonquin wolves, even though small relative to gray wolves, are efficient moose predators, able to kill adults in both deep and shallow snow. There is no mistaking a wolf kill; the forest around the moose carcass is smashed up, although on one occasion we tracked a pack to a kill in deep snow where the moose was unable to extricate its hooves to defend itself and was killed on the spot. Moose, like deer, make up only one-third of wolf diets, reducing the impact of predation. Even calves are only lightly taken, representing 15 percent of summer diets.

Then there is beaver, abundant yet limited in availability in winter by ice and in summer by the attractiveness of offshore aquatic vegetation. Based upon government aerial surveys and our ground observations, beaver num-

bers have stayed relatively constant in the eastern half of Algonquin, dropping only in the west. Few studies have shown a significant reduction in a
regional beaver population due to wolf predation. Beaver declined on Isle
Royale in Lake Superior when wolf densities were at an all-time high and
moose had declined to a 12-year low. No deer live on the island. For most
wolf populations, beaver is a "filler," taken in large numbers only when
deer, their preferred food, are low. At Algonquin, preference for deer is
shown by wolf movements—shifting seasonally as deer do and concentrating with them in winter yards—and eating a higher proportion of them as
they decline instead of dropping their use proportionately.

As a broad generalization, wolf abundance tends to correlate with prey
abundance, or more accurately, with prey biomass, just as one might expect.
This relationship was illustrated first by Wisconsin biologist Lloyd Keith,
then by Massachusetts biologist Todd Fuller, in graphs that included results
from studies across North America. But while prey abundance may influence general wolf population levels, other environmental conditions can
set numbers even lower or be the cause of annual change. That is the case
at Algonquin. Over the twelve years of our study, in contrast with the wolf
decline, moose numbers first decreased, then increased, then decreased,
ending at about the same level as when our study began. So did deer, following a reverse sequence to about their same level as at the beginning.
That leaves us unable to account for the wolf decline by changes in prey
abundance.

In the northeastern United States, prey abundance will undoubtedly set
at least a potential size for any recovered wolf population. Over most of the
Northeast, deer numbers exceed those currently in Algonquin Park. In the
Adirondacks, deer numbers are close to twice that of Algonquin, in west-
central Maine about five times, and in parts of Vermont ten times as abundant. Moose reach or exceed their Algonquin density in west-central and
northern Maine, and northern New Hampshire. In those areas of highest
prey density, wolf numbers theoretically could reach twice their current
density in Algonquin Park, exceeding three per 100 square kilometers as
they once did in Algonquin. As this figure suggests, the relationship between wolf and prey numbers is not simply proportional, one-to-one. Cor-
roborating our figure are other estimates of potential wolf density made by
D. Mladenoff and T. Stickley, and by Todd Fuller based on extrapolating
data from the Great Lakes states.

But as mentioned, prey is not the most immediate cause of the decline in Algonquin wolves. One important characteristic of the population is a low level of annual recruitment; low numbers of pups and/or low pup survival. A young animal is not considered a recruit until it reaches breeding age, for wolves usually two or even three years of age. The level of recruitment is important because a population cannot experience more than that amount of annual mortality without declining. In Algonquin, each year, the average of yearlings in the population in late winter is approximately 21 percent, not yet even technically recruits. This figure is low compared to others in the wolf literature, especially those describing exploited populations. It leaves the Algonquin wolves especially vulnerable.

We can only speculate on the reasons for this low recruitment, because we do not radio-collar pups for fear of disturbance. One possibility is disease. Veterinarian Ian Barker at the University of Guelph has screened the blood serum of animals we have handled for antibodies that defend against various viruses. His findings show that canine parvovirus is widespread throughout the population. This disease can kill pups between eight and twelve weeks of age after they lose a maternally endowed immunity and before they can experience low levels of the virus to build up protective antibodies. But against this explanation is the evidence that Algonquin wolves were experiencing similar low levels of recruitment back in the early 1960s. Back then, canine parvovirus was unknown anywhere. It evolved from a feline parvovirus only in the 1970s.

Another possible cause of low recruitment is poor pup survival due to inexperienced adults. In the Algonquin population, a two-year-old female breeding for the first time will have a chance to breed, on average, only once or twice more. Life expectancy is low. Many studies of social canids and primates have demonstrated the importance of experience in the care and raising of young. Young animals face perils that require vigilance from adults including accidental abandonment when the pack moves from its den site to various rendezvous sites, defense against bears, and guidance away from humans and vehicles—a well-founded fear. As well, parents and pack members must provision the young with food and teach, by example, to hunt and kill. While it is difficult to gather empirical data on such social traits, our evidence indicates that pup losses are high. Late-summer litter sizes are in the order of one, two, or three, not four, five, or six as more commonly reported in other studies. We have noticed differences in the fidelity of mothers to den and rendezvous sites that may be interpreted as differences in caring or differences in the mother's hunting skills or those of her pack. Once we recorded an abandoned pup that starved in a rendezvous site after the pack moved on.

Low productivity is detrimental to a population only if annual mortality exceeds it. In most years of our study, it has. The worst single years were 1987/88 (57 percent mortality), 1992/93 (48 percent), 1993/94 and again in 1998/99 (39 percent). Wolves have died from being pummeled by moose hooves, or from one swift kick by a deer, or attack by other wolf packs, or rabies, or starvation.

Cases where moose or deer kill wolves can be thought of as accidents, and they are rare, only occurring once in our study. Upon autopsy, though, we have found many skeletal injuries such as broken ribs and even broken legs that have healed. Aggression from other packs has accounted for only four deaths, surprisingly low because trespass and close encounters between packs are very frequent in the Algonquin population. A rabies outbreak swept the population in 1990/91 killing six of eighteen radio-collared wolves, a significant source of mortality in the population, unexpected because rabies in wolves is rare south of the Arctic. We searched all records in Ontario and found fifteen cases between 1960 and 1994. In contrast, in the Great Lakes states, rabies has been recorded only once. Different strains of rabies have been involved; skunk in the Great Lakes states and red fox in Ontario, but we do not know the significance of that fact. Vaccine distributed out of airplane windows in balls of meat is quickly making rabies a thing of the past in Ontario, except for the threat of a raccoon strain crossing the border from New York State.

Our few cases of starvation all occurred in late summer or early autumn. By that time, young deer and moose are fast on the hoof, and beaver are still relatively unavailable out in the ponds. Starvation is partially linked in our data to particular social situations, such as being ostracized from the pack.

Wolf packs are predominately, but not entirely, family units with genetic bonds that underpin cooperative behavior. At one level, packs operate as biological units with members dependent upon coordinated and combined skills as hunters, as caregivers to the young, and as territorial defenders. Because of these group traits, to some extent all pack members share a common fate. They compete for land with other packs. Algonquin packs average about 150 square kilometers each. Any seemingly unused land is only temporary, in interstices between territories, or in vacant territories as a result of the death of entire packs, an unfortunately common event. The success of packs in this competitive environment is dependent upon all its members. It is axiomatic that the most fit packs will survive best and likely leave the most offspring.

At another level, however, pack members compete with one another for food and mates. Normal natural selection takes place among individuals. Situations that work to select individual wolves include social dynamics that result in outcasts due to either injury or old age. With accompanying disabilities, these animals in our study eventually starved. Small wolves, overall, seem to die sooner than large ones; we are examining our data. If that conclusion bears out, the reason might involve social subordination within packs that might limit access to food or cause stress.

These natural sources of mortality, however, do not predominate at Algonquin Park, or almost any place where wolf research is ongoing: Yellowstone National Park in Wyoming, Alligator River National Wildlife Reserve in North Carolina, Pukaskwa National Park in Ontario, Banff National Park in Alberta, Laurentide Provincial Park in Quebec. In all of them, the main cause of wolf deaths is humans. In Algonquin, like the others, human killing exceeds all sources of death put together, regardless of "protected" status. We find no evidence that this killing is "compensatory." The wolves that were killed were going to die of some other cause anyway, as is sometimes the case in wildlife populations, especially with young, highly susceptible animals. In our case, statistically, human-caused mortality draws down the population.

THE ECOLOGY OF WOLF-HUMAN RELATIONSHIPS

Studying the population ecology of wolves anywhere unfortunately involves assessing the why and wherefore of human killing. Detailed habitat analyses such as have taken place in the Northeast are necessary but not sufficient. Good habitat for these animals is there, but so are humans.

Some species, especially the extinction-prone, need specially protected lands and laws to sustain them. These species include habitat specialists, species with low birth rates, species that live at low density, large-bodied species, summit predators, exploited species, and persecuted species. The wolf is all but the first of these. It needs protection.

The parks of the world are failing to protect the large carnivores of the world because of their extensive space demands and increasing human densities in surrounding lands. Managers in all but one Canadian National Park with wolves, including parks throughout the arctic, recently listed wolf transboundary problems as a major park issue. Snowmobiles and airplanes have extinguished remoteness. The size and persistence of any park wolf population appear to depend on the degree of exploitation immediately outside.

The territories of approximately half the wolf packs of Algonquin Park overlap park boundaries. Fully half the park lies within 10 kilometers of a boundary, a geometric reality of an approximate rectangle like Algonquin. The diameter of wolf pack territories exceeds that, variable but averaging 14 kilometers. Consequently, only the core-dwelling half of the population holds lands completely within the park. When a wolf crosses the park boundary, unmarked and invisible and impossible for it to comprehend, protection ends.

Yet a second circumstance draws wolves out of Algonquin Park: the annual deer migration. Each year between mid-November and late-January, all, or almost all the deer occupying the eastern 2,700 square kilometers migrate to one major or two satellite deer wintering areas a few kilometers east or southeast of the park. Snowfall there is lower, and the broken forest-farm interface creates favorable browse. Cedar swamps, rare in the park, provide good cover, as do pine plantations. As well, people put out hay as supplemental deer food and to see them feeding close to their houses and barns. Small towns and villages dot the 180 square kilometers of deer yard, although it is about 80 percent forested, and the land is gridded with roads. Most of the wolf population, spread each summer across the same 2,700 square kilometers of deer range in the park, follows the deer out, too.

In Ontario, trappers are restrained by no quota on wolves and have no requirement to report their kills. Landowners or people holding a small game license can kill wolves year-round. No other jurisdiction in Canada, and few in the world, has such appallingly out-of-date, exploitive wolf management policies. Despite considerable public objection, backcountry politicians and senior government policy makers apparently selected as status-quo clones appear intent on keeping it that way. A government-sponsored five-year policy review published in 1998 recommended game status for wolves with area-specific harvest levels and monitoring. However, the report gathers dust despite the prodding of both national and provincial citizen conservation organizations.

Because Algonquin wolf packs exhibit the most extensive off-territory movements recorded for deer range, we have studied the phenomenon in detail. Elsewhere, one or two packs may concentrate their winter activities where deer yard up, but extensive movements of many packs away from their summer range has been described only three times, where wolves follow migratory bison or caribou in the Northwest Territories or Alaska.

One reason for the off-territory movements at Algonquin appears to be that prior to 1993, resident canids living year-round in the deer yard likely were being eliminated by humans, leaving undefended land for transients to occupy. There is a lesson here. Before 1993, the transient Algonquin Park wolf packs used the entire deer yard. Then in 1993, the Ontario government implemented a wolf-killing ban in the deer yard, effective each winter while Algonquin Park wolves are in residence. After that, the park packs stopped using most of its core. Since the winter of 1995/96, graduate student John Pisapio has found two resident packs in that core. By radio-tracking them, he determined that they were able to restrict the movements of the transients; even a pack of only three resident animals was successful against transitory wolves in packs as large as nine. Ironically, the people who previously killed wolves in the deer yard hoping to increase deer for their own hunt, may have killed resident wolves and thereby invited a greater number of transient wolves in, resulting in more predation than if they had left the wolves alone.

This relationship between resident and migratory wolf packs illustrates the unexpected. Any evaluation of the Northeast for wolf recovery can count on that, too. Wolves, with their great behavioral plasticity, are far from predictable.

The conditions at Algonquin, where the fate of a large carnivore population depends largely upon what happens outside the park, are often described in journals of conservation biology. Recommendations repeatedly are made that park management must not end at park boundaries. It would be risky, for example, to protect wolves in the Forest Reserve portion of Adirondack State Park but ignore possible wolf movements and mortality beyond. As self-evident as this is, often the mandates and responsibilities of government agencies form impenetrable walls. Many Algonquin Park wolves die because "Ontario Parks" is a separate agency from the rest of the Ministry of Natural Resources. Its authority ends abruptly at park boundaries. Nothing, it seems, is guarded so strongly as jurisdictional turf, regardless of conservation consequences.

So, for the past twelve years, we have tracked an objectionable number of wolves to peoples' houses and barns, and trucked carcasses back south for necropsy at the University of Guelph's Veterinarian College. Many of these wolves we have known well, like Jocko 10, others more fleetingly. We have identified the following facts about this killing:

- Some has been done by commercial trappers but more by people who hate wolves. They bait and shoot or snare them, then leave their carcasses in the bush or discard them unskinned.
- Most people who kill wolves are middle-aged or older.
- Even at the height of the killing, only a few individuals were involved, although large, organized snowmobile shoots have occurred.
- Poison (strychnine) was used only once to our knowledge. The person was not apprehended.
- When the government instituted the wolf-killing ban, local opposition was heated, with raucous town-hall meetings. A wolf head and smashed radio collar appeared on a telephone pole across from a large church as a sign of protest. Hate letters against both wolves and us filled the local newspapers. Still, a local editor who periodically published our progress reports told us that only a few people were behind the protest; many more were for wolves, or indifferent, but would not speak out because of community and family relationships.
- Despite the protest and statements by people that they would flout the law, most people abided by it. Most trappers were cooperative, recognizing the incongruity between unlimited and unregulated harvest of wolves versus sustainable harvesting regulations that they had to follow for all other species.
- After the wolf-killing ban, the wolf population has continued its decline to an all-study low. Most of the known killing now occurs in townships adjacent to the park other than the ones covered by the ban. Clearly the ban covers too little of the adjacent land.
- In trying to rectify this situation through discussion and reason, we have found many local people to be more open-minded than many government officials. Some influential bureaucrats trained to sidestep issues have denied that wolf conservation is a problem at Algonquin at all, or anywhere else in the province. The Ontario government has not collected any data on wolves for thirty years and has no practicing wolf biologist.

Consequently, of overriding significance in wolf-human relationships is not what goes on out in the bush, or even the hunt camps or fur auctions, but in the boardrooms and government offices. It takes civil servants with conviction to encourage their governments to tackle difficult issues like wolf management. The system often does not reward those who draw fire. Conservation policy jerks along unevenly, waiting for individuals with the right combination of ethics, integrity, and power.

But, even bolstered by conservation-minded biologists like those involved in current wolf restoration programs in the United States, governments cannot act alone. Public opinion is crucial. Over the long haul, despite lags between public demands for change and government inaction, democratic governments deliver what the public wants—and deserves.

Not all species recovery issues call for the same approaches. Friendly committees, discussions, and working groups work fine for the Furbish's lousewort or the Fowler's toad, both endangered species in Canada and both totally benign to human welfare. For wolf conservation, however, it is impossible to find successful examples that used this approach. Wolf management invariably generates heated public exchange. It is sure to draw out strong, even violent emotions. It highlights conflicting worldviews about who and what "resources" are for. The result is political wrangling.

Wolf biologist David Mech ended his famous 1970 book *The Wolf* with the statement, "If the wolf is to survive, the wolf haters must be outnumbered. They must be outshouted, outfinanced, and outvoted." While that may not sound very democratic, neither is government tiptoeing around difficult issues. Lobbying is part of Western democracy, hopefully with the facts straight and the unanswered questions recognized as such.

Regardless of how decisions are made, the eastern forest wolf will not survive indefinitely by default. Its future depends upon legislation and planning. Like all the water flowing into Everglades National Park, eastern wolves are under our boardroom control.

That is, except for the "coyote connection."

THE COYOTE CONNECTION

A less obvious and more insidious threat than population drawdown to Algonquin wolves and red wolves is posed by their little relative, the coyote. Coyotes are gaining ecological dominance by hybridization: coyote gene introgression leading to gene swamping and ultimately the genetic transformation of the species. The process is well underway in both Ontario and North Carolina. It greatly complicates wolf recovery in the northeastern United States, because eastern coyotes live there, too.

Spring had brought the azaleas into bloom early at Virginia Beach on the Atlantic coast in 1999, but we didn't have time to enjoy them much. We only saw them on the drive to and from the airport. For four days, along with forty other biologists, we were cloistered away at a convention center. Under the auspices of the World Conservation Union's "Conservation Breeding Specialist Group," we were taking part in a "population and habitat viability analysis" (PHVA). Drs. Ulysses Seal and Phil Miller and their colleagues travel worldwide staging these workshops for the benefit of populations and species about to enter the "extinction vortex." At some critically low population level the inescapable vortex drains away the last of a species. The workshops are designed to plug the drain with a set of recommendations based upon the best available expertise. The issue in Virginia was in sharp focus: How do you stop the hybridization of red wolves with coyotes in the Alligator River National Wildlife Refuge?

The red wolf—North American–evolved wolf—and the coyote may have danced together, sharing genes, ever since the two split into distinct or quasi-distinct species between 200,000 and 300,000 years ago, as described by Brad White and colleagues based upon mammalian rates of genetic molecular divergence. This proposed history makes more sense than a previous one derived from skull measurements alone, which suggested divergence more than a million years ago. The more recent divergence explains why the two species can interbreed; 300,000 years is not long in evolutionary terms.

The condition of their divergence is totally unknown. It may have occurred in the pre-Illinoian interglacial period. Prodigious amounts of water had to have melted from a previous glaciation. Possibly a great prehistoric river, an early Mississippi, formed a geographical barrier that split the population and provided the genetic isolation necessary for speciation to occur. All we know is that at the beginning of European recorded history, coyotes lived primarily on the open plains of central United States and red wolves occupied the southeastern forests. Their ranges overlapped little, doing so only in eastern Oklahoma and eastern Texas.

A species is not easy to define. If two "species" live in different places, but together could and would interbreed, are they only one species? Ornithologists have puzzled over this question, splitting then rejoining Bullock's and northern oriole, slate-colored and Oregon junco, myrtle, and Audubon's warbler. In response to human alteration of landscapes, many species have extended their ranges into those of closely related ones. These species cause taxonomic chaos.

Since European colonization of North America, coyotes have expanded in all directions from their original prairie range. Good evidence links their expansion with the decline of wolves, gray and red, not necessarily complete extirpation, but at least fragmentation. In the East, coyotes hybridized with red wolves. Coyotes did not, then or now, hybridize with gray wolves, if you accept Brad White's hypothesis over Bob Wayne's, as described earlier. Gray wolf and coyote occupy the same ranges throughout much of western North America—Banff, Jasper, Yellowstone, Yukon. At Yellowstone, wolves routinely kill unwary coyotes that come scavenging on their kills.

In the 1930s, red wolves from Arkansas weighed the same as Algonquin wolves do today. But by the late 1960s, Arkansas red wolves were smaller —morphologic evidence of the influence of coyote genes. All biologists studying the fate of red wolves agree that coyote gene swamping was eliminating the species. Reports of full-bodied red wolves declined as coyotes moved east, eventually crossing the Mississippi and invading the extreme southeastern states. The last red wolves holed up in Louisiana bayous, then they could be found no longer. Soon after, they were declared extinct in the wild.

Fortunately, some full-bodied red wolves had been put in captivity in the 1980s and were released first at a number of island sites off the eastern coast of the United States, then at the Alligator River National Wildlife Refuge, later in the Great Smokies National Park in Tennessee. At the time of our PHVA, the Smokies release had failed largely due to dispersion into areas of insufficient food. The only hope for a self-sustaining population lay at Alligator River. There, red wolves numbered approximately seventy, a success built upon repeated releases as well as wolf protection, and considerable efforts to achieve public acceptance. Prospects for long-term sustainability seemed within grasp.

But then coyotes expanded in northeastern North Carolina. A principal reason for choosing the Alligator River Refuge had been its near absence of coyotes and its insularity, which reduced the probability of their invasion. But it was only a matter of time; good farmland coyote habitat abuts the refuge to the west. At the time of the PHVA, an estimated 20 percent of red wolf matings were with coyotes.

Our study group explored all possible options short of fencing off the reserve, seen as an admission of defeat. No alternative coyote-free recovery sites in the eastern United States were known. The final recommendations were not comforting: eliminate or sterilize all coyotes found on the refuge for three to five years and reintroduce red wolves from captivity to fill any territorial vacancies in the hope that a wolf fortress develops that will prevent further coyote invasion.

Nobody was really pleased with the management prescription. The world has changed around the red wolf. How far do you go to perpetuate a species? After five years, if the plan were not successful, the problem would be reevaluated with the options of fencing—the end of free-living animals—or simply letting the coyote genes take over.

And what of other red wolves—"*lycaon* wolves" elsewhere—the White hypothesis? Extinct in the northeastern United States and the Canadian Atlantic provinces; alive and "protected" in Algonquin Park; integrated in some undefined way with *lupus* and protected in the Great Lakes states; integrated and unprotected in parts of eastern Manitoba, western Quebec, and to the east and north of Algonquin Park; gone south of Algonquin down the Frontenac Axis.

That axis has often been described as the "Algonquin to Adirondack" corridor of potential natural wolf recovery. It is too late. With graduate student Hilary Sears and other students, we have examined trapper-killed carcasses, measured bodies and skulls, taped howls, collected scats, described and quantified the landscape, and radio-collared three packs in the wildest landscape at its northern end near Algonquin Park. Brad White, Paul Wilson, Sonya Grewal, and colleagues have performed genetics testing. Results: hybrids. To the west of Algonquin Park: hybrids too, albeit with closer similarities to Algonquin wolves. Territory sizes of Frontenac Axis animals are in the order of 35 to 40 square kilometers: coyote-like, not wolf-like ranges. Groundhogs, rabbits, and vegetation, coyote-like foods rarely eaten by Algonquin wolves, show up in scats. Howls are nasal and high-pitched. Feet are small.

In recent years, scientists have puzzled over these animals, suggesting hybridization, calling one group a new subspecies, the "Tweed wolf," separating an Algonquin-type gray wolf from a boreal-type gray wolf.

Now, with confirmation that these Frontenac animals are genetically more coyote-like than wolf-like, we can piece together what has happened. Coyotes invaded southern Ontario early in the 1900s, concurrently with increasing human populations and attendant wolf persecution. Wolves gave way but at the same time interbred with the invaders. The same phenomenon of coyote gene swamping that happened to the red wolf in the southeastern United States has happened in southern Ontario.

Careful scrutiny of landscape type, road density, and canid characteristics on the Frontenac Axis show that while wolf-likeness is greatest in closed forests, most important is road density. Where human access is greatest, providing some measure of opportunity for killing of wolves, wolves have given way most completely to hybrids. The system has flipped. The adaptable, more fecund, less space-demanding coyote—or more accurately, coyote-wolf hybrid—has taken over.

The relationship between road densities and the persistence of wolf populations has received a lot of attention recently with studies in the upper Great Lakes states, western Ontario, and Alaska. Roads that fragment blocks of public land reduce wolf populations by providing access for hunters and trappers, causing road deaths and increasing areas that wolves simply avoid. Wolf populations drop where all-season two-wheel-drive road densities are greater than approximately 0.6 kilometers per square kilometer, but wolves in the core of their territories are even more sensitive, tolerating less than half that density. These figures were derived mostly from jurisdictions with full wolf protection. Even in protected places, human access invariably results in killing.

On the Frontenac Axis, three conditions were different: we included lower-quality roads and trails suitable for four-wheel-drive and all-terrain vehicles; we related road density to hybridization, not just to population reduction; and we were studying unprotected populations. Hybridization showed up on all six blocks of land studied, influencing body morphology, food habits, and genetics even on the wildest lands with no all-season roads at all and a density of just 0.3 kilometers per square kilometer of four-wheel-drive roads and trails. Such old logging roads and trails service a multitude of hunt camps and provide access for trapping.

A number of biologists have drawn maps of potential wolf habitat in the Northeast. Not considering four-wheel-drive roads and trails weakens their results. So, too, does the lack of attention to the threat of hybridization that might occur at a lower threshold than obvious population reduction, which we cannot yet measure or predict. This is knife-edge ecological fragility, difficult to predict, but important when the persistence of *lycaon/ rufus* in all its remaining eastern haunts may depend upon it.

Even the protected wolves inside Algonquin Park have been hybridized, but to a lesser extent. Protection appears to have just slowed the process. Sonya Grewal reports that coyote mitochondrial DNA haplotypes have changed from 18 percent to 48 percent in the last thirty years. Paralleling these genetics results, we find that skull sizes of Algonquin wolves are slightly smaller than they were thirty years ago. Although average body weights are similar, females weighing less than 50 pounds have increased by close to 20 percent. We see a coyote-like high level of tolerance between packs, more than expected for wolves, especially in the deer yard, but we do not know if hybridization or the influence of abundant prey is involved.

While the increase in coyote-like characteristics may be influenced to some extent by drift of the original genetic material, continued hybrid immigration into the park is clearly taking place. We have recorded trans-boundary movements deep into the park by a few hybrids, some electing to stay. As well, wolves, or even packs, pick up recruits during the breeding season when they are occupying hybrid lands outside the park, a phenomenon that occurred so often in the major deer yard that former Ph.D. student Graham Forbes called the deer yard a "singles bar." Also, small, coyote-like animals tend to show up in the park disproportionately in fractured packs or where there are territorial holes. Human killing has caused plenty of territorial holes; it often follows a different pattern than natural killing. A wolf pack goes down a snare line some starry night, and most of the pack has strangled by dawn. Recently, an Algonquin pack swung through a military reserve adjacent to the park and all were shot.

Given these movements, we expect the effects of gene swamping to be even more apparent than they are. Something besides protected status may be working to slow the hybridization process in Algonquin Park wolves. We can only speculate on what that might be. Possibly sexual selection works against small animals, that is, given a choice, wolves may choose large mates. Dominance within wolf packs often, but not always, goes to large animals.

While the process of gene swamping has been slower in Algonquin than outside, the end result, over time, is predictably the same. Conservation groups, and we the researchers, argue for a wolf protection zone around the park to increase the strength of the wolf fortress, in hopes that the park will be less permeable to hybrids—the same strategy that is being employed for the red wolf. Except for the wolf-killing ban in the deer yard, so far we have argued in vain.

Just as wolf populations in the East are becoming more coyote-like through hybridization, coyote populations are becoming more wolf-like. The phenomenon appears to be an example of ecological convergence, or more likely homogenization of a gene pool too recently diverged to be permanent. Coyotes invaded southern Ontario in the early 1900s to fill a wolf void on forested lands recently scalped for agriculture, then moved on to the north side of the St. Lawrence River by 1935, and to southern Quebec a decade later. About that time, too, they began to fill a forty- to sixty-year wolf void in New York State. Earliest reports there seemed to be of coydogs (coyote-dog hybrids) and involve escapees or unauthorized releases, but by the 1960s coydogs had largely disappeared and the central Adirondacks held animals more clearly identifiable as eastern coyotes. Since then, these animals have moved on to the north and east, filling the rest of the states and eastern Canadian provinces, progressing in the past decade even to Newfoundland. That was just their eastward expansion; they moved west, too, and north into the Yukon and British Columbia. Everywhere they went, either the wolf had been extirpated or humans had heavily impacted their populations.

As they spread northeast, they became larger, "little wolves," to overlap in size with the small end of the wolf-scale, but only in the Northeast. That geographic fact provides a clue to at least one of the causes. In the East, they would have run across *lycaon* wolves, their progenitors according to the White theory. In the West, they met *lupus*, with whom they show no evidence of hybridization. Brad White and his colleagues have found widespread *lycaon* alleles in a sample of trapper-killed coyotes from the Adirondacks, and even an "Algonquin-specific allele" that so far has turned up in no other wolf population. The presence of that allele suggests that New York coyotes may have picked up at least some of their wolf genetic material in Ontario before expanding their range to the south. Maybe there were a few wolf hangouts in New York, too—nobody knows for certain.

Anyhow, eastern coyotes are predominantly hybrids, more coyote-like than wolf-like, even more so than the hybrids of the Frontenac Axis. The occasional very large animal that shows up in the Northeast is either an extremely large hybrid carrying more wolf alleles than a smaller animal, or may even be a wolf, as were two animals recently killed in northern Maine, confirmed by genetic analysis. Possibly these animals migrated from southern Quebec.

Nevertheless, coyote-wolf hybrids are common today throughout the northeastern United States, and they push the vacant wolf niche. Instead of concentrating on microtines and rabbits, they have expanded their food niche to include white-tailed deer and beaver. Beaver were absent in coyote diets in the Adirondacks until the late 1980s, but when Bob Chambers conducted studies there in 1996 and 1997, he found beaver to constitute 12 to 15 percent of summer diets. There was no accompanying increase in beaver abundance, possibly even a decline, indicating that the hybrids must have been deliberately selecting them. Deer populations had remained low and steady. On the Gaspé Peninsula in Quebec, beaver increased in June diets of coyotes between 1988 and 1991 despite a drop in beaver abundance, again showing selection. In contrast, smaller western coyotes, with jaws limited in grasping power, have never been reported as significant predators on beaver.

White-tailed deer, another traditional wolf food, is common in the diets of eastern coyotes, too. Gary Brundige described an increasing proportion of deer in coyote diets of the central Adirondacks between the late 1950s and late 1970s, despite a significant decline in the deer population. This increase in predation on deer, he felt, reflected the development of a food preference and the formation of larger packs as the coyote population increased. Coyotes tended to hunt in groups when pursuing deer, and were most successful when snow was deep or crusted. Nevertheless, adult deer made up more than half the summer diet. Similarly, biologists studying eastern coyotes in western and southeastern Maine reported deer as most important, followed by hare. On the Gaspé Peninsula, moose were important but were likely scavenged.

Some biologists have speculated that the larger size of the eastern coyote is due to natural selection for more efficient predation on larger prey, specifically deer. Selection based upon the size of prey, versus selection to handle deep snows, versus nutrition—these three possibilities cannot be separated easily. Canid adaptability in eastern North America well may involve them all.

Not only have coyotes converged on wolves morphologically, but behaviorally, too. Examples abound of eastern coyotes forming substantial packs. Again, this tendency may be selected to improve hunting efficiency for deer. It makes no sense for a pack of canids to stick together to hunt field mice. But similar large packs have been reported among western coyotes where they have not been exploited in California, Yellowstone, and Arizona. Large packs may be beneficial in pup rearing and territorial and carcass defense. Some people wonder if the supposed coyote norm of predominately small packs, early dispersal of young, and less rigorous territoriality than wolves, extracted from the great bulk of western coyotes studies, misrepresents truly natural coyote behavior because almost all studies have been of exploited populations. We see similar social disintegration in exploited wolf populations, as well.

The land demands of eastern coyotes are quite variable but also converge upon those of wolves. Western coyote territories are in the order of 10 to 30 square kilometers, whereas those of eastern coyotes have been recorded at 45 square kilometers in Vermont and up to 112 square kilometers in the Adirondacks. Brundige attributed the latter figure, a big jump from other studies, to a relatively low deer density. Notwithstanding the link between prey density and territory size in both wolves and coyotes, packs of eastern coyotes tend to use more land than western coyotes.

To a large degree, then, the coyote hybrids in the Northeast are functional, or slightly dysfunctional wolves, behaving like wolves, looking like wolves, eating wolf foods—filling their niche. Not completely—they are not predators of moose, a significant difference especially in Maine. Their use of beaver is considerably less than that of wolves, except in the Adirondacks. Rabbits notice the difference between wolf and coyote, being an important food only for the latter. The rigidity of coyote territoriality in general seems to be less than wolf; evidence is provided by a 32 percent overlap between coyote packs and seasonal shifts in the Adirondacks, although there, packs did not travel to winter deer yards as Algonquin wolves do. Differences in territoriality may be the result of local ecological conditions such as the presence of resident packs. Coyotes reportedly hunt over smaller areas each day and season, even where their all-year territories were large in the Adirondacks. In contrast, wolves tend to constantly cross and recross their entire territories. Coyotes form denser populations, averaging around one per 10 square kilometers in the eastern forests, while a dense wolf population is one-third as great. Coyotes breed at an earlier age, and that may enhance their density, too. Even though forming denser populations, coyotes eat less meat per animal per day, and in the summer, more vegetation, so their impact on deer is tempered in that way. And coyotes may have a greater predilection for eating livestock—most of Ontario's depredation involving wild canids is done by them. Ecological replacement is never exact.

But replacement in the Northeast is close enough to suspect, under the broad umbrella of canid behavioral plasticity, that any wolves released or finding their way there would be accepted into the canid society that already exists. The only way to increase the odds of preventing hybridization would be to reintroduce wolves into a large block of land, at least as big as the 7,500-square-kilometer Algonquin Park, which has been rendered coyoteless. Even if that were feasible, it would likely take continued coyote removal and strict wolf protection to bolster the hopes that a wolf "fortress" would build up. Wolf re-establishment would be a possible, but providential outcome.

Instead of re-establishing a wolf population free of hybrid influence, the more likely result of reintroduction would be a shift toward wolfness in the existing coyote-dominated gene pools. Recolonizing wolves will already carry some coyote alleles, wherever they come from, because all *lycaon* do. In the northeastern United States, they will undoubtedly hybridize with coyotes that already carry some wolf alleles. However, if body size is important in mate selection, as is repeatedly suggested for the wolf, then wolves, being larger, would have more than a random chance of not simply degenerating to some hybrid average. If it is also true that an intact wolf population can repel coyotes, then chances beyond random increase some more.

Such a goal for species recovery—gene pool shift rather than species reintroduction—is unique, never having been attempted or even contemplated before. The Northeast offers a grand opportunity to try it.

MADE-BY-HUMANS WOLVES

Like it or not, much of nature in the Northeast, indeed in much of North America, and even in the world, is increasingly under human control. Subtly, natural selection gives way to human selection as a force continuing to shape living things. "Wildness," child of nature-forged life-on-Earth, retreats as the new selective spotlight falls on us.

Conservation is not concerned much with quality. With backs to the wall, mere species survival is of singular importance. But what sort of species are we saving—those fashioned by natural selection or those that adapt to our remodeled world? If the former, what about the need to maintain the natural environmental crucibles that forged them?

When the dominant cause of mortality in a wolf population becomes human killing, human selection has taken over. Then, selection occurs for the skills of staying out of neck snares or keeping out of sight to avoid being shot rather than for group cooperation, hunting prowess, speed, or endurance. That is the situation for wolves over much of their range. Will the measure of success for any reintroduction of wolves into the Northeast be mere population persistence despite exploitation? Will we be content simply if their reproductive capacity exceeds our butchery?

The most important long-term effect we have upon wolves could be upon their sociality, the adaptive key to their success. They are among the most highly evolved social mammals in the world. What level of persecution begins to erode their complex social traits? In Algonquin Park, we are concerned that a loss of pups may be due to inexperienced adults, as discussed. We see elevated levels of dispersal and adoption, and little traditional use of den or rendezvous sites. Young wolves must learn most of their hunting and social skills through observation during a long period of dependency. Pack fitness suffers when team leaders are no longer present; while a pack may be the same size, it is like a hockey team without its stars. In subtle ways, we risk reconfiguring the wolf.

This problem is not confined to wolves, of course. Cheetahs reportedly hunt more at night in heavily visited eastern African parks, leaving their young more susceptible to night-hunting hyenas. Bears adjust their social behavior to forage at garbage dumps. Dingoes space themselves differently to take advantage of bores for watering cattle. Wherever humans alter the environment, species have to shape up or face an uncertain future.

We may tend to think of natural selection as operating over vast periods of time, but that confuses its role in speciation—the formation of new species—with "sequential evolution," the adjustment of gene pools to more immediate environmental change. The latter can happen quickly, especially when a population suffers heavy mortality. Well-known "industrial melanism" happened to moths in Britain in just a few decades, where pollutants blackened smokestacks and avian predators picked off individual white moths more easily than melanistic ones. Finches on the Galapagos change beak characteristics from year to year as El Niño events favor different sizes of seeds.

The extent to which selection on wolves is predominately human or natural depends upon protection. Protection has been the big difference between the successful expansion of wolves into Wisconsin and Michigan, where they have it, and wolf decline suffered outside Algonquin Park, where they do not. Even though we do not know if the expanding populations are dominated by *lupus* or *lycaon*, and so may or may not hybridize with coyotes according to the White hypothesis, still they have had protection. That might have been enough for them to repel coyotes, even if they are *lycaon*. They have had the advantage of expanding from a front of healthy wolf populations in Minnesota. Unfortunately, they lack that opportunity in the Northeast.

In any planned reintroduction of wolves into the Northeast, even with protection, we still have one more important choice. One option is to put them into logged-over and human-settled landscapes. Is that worth it; is it even ethical, considering that they invariably will suffer from human persecution? The other option is to put wolves where the full crucible of nature acts on them, their prey, and the forest associations: in ecosystems as free as possible from human influence. That is what national parks are for, such as the proposed Maine Woods National Park. Bringing wolves to such a sanctuary from *lycaon* lands that may exist north or east of Algonquin Park where they and the ecosystem are open to human disruption, would truly be doing something for wolves, not just for us.

Reintroduction into such a place, a new national park, would likely preclude a slaughter of all the (hybrid) eastern coyotes there now, by park policy. Short of the unlikely possibility of wolves replacing the existing hybrids, as discussed, success in wolf restoration would have to be recognized as a shift in the gene pool that makes the hybrids exhibit characteristics more typical of a protected wolf population: an increase in wolf-like territory sizes, low levels of dispersal and adoption; high degrees of social cohesion and territoriality; greater use of beaver especially where deer numbers are low; lower percent use of rabbits and fruit; and perhaps most importantly, greater predation on moose. Lack of successful predation on moose is the hybrids' biggest "failing" in filling the wolf niche. Ecologically, this failing justifies the efforts of planned wolf recovery, especially in Maine and New Hampshire, with their large moose populations.

These qualitative characteristics can provide real, on-the-ground measures of the presence of wolves, evidence that the animals are filling the wolf niche. We may never be able to distinguish wolf from coyote from hybrid based upon genetics and morphology, and may always be tripping over the definition of a species. But based on ecological role, a definable wolf might again occupy the Northeast.

If our research in Algonquin has taught us anything, it is that wolves are unpredictable. We should have expected that; few species are programmed with such freedom of choice. Generalist among generalists, most adaptable among the adaptable—except for what we humans can do to them. So it is time to write a disclaimer. They might surprise us again and, despite all the foregoing discussion, put the run on the hybrids. I doubt it, but it is not beyond the bounds of possibility. Hybridization is a fickle biological trait. Barred owls that have recently moved west to invade coastal forests both chase out spotted owls and breed with them. Wolves and coyotes are not the only species collaborating to confuse us and complicate conservation.

But then, what about the rest of the land outside any parks or recovery zones? Maybe over time wolves will get there, too, but maybe not. Why not accept the hybrid? It has shown itself to be an adaptive genotype in a human-dominated, logged, and lived-in environment. Fortunately, eastern wolves were adaptable enough to hybridize, leaving many of their genes still out there even though they have gone. Their genes managed to invade the invaders, retrofitting them so coyotes show more wolfness than they did on the prairies where they came from. Why slight the hybrid that has done such an admirable job at pinch-hitting for the wolf, especially in the Adirondacks, where they most closely fill that niche? Instead, why not be grateful to hear its howls and see its footprints in the mud? Hybridization has always been a major evolutionary force in creating the world's biodiversity.

A broad public will have to decide what to do, based on perceptions and values. Science can only supply some estimation of the consequences out there in the forest. If the experiment is worth trying, it will be because people want to. Alternatively, the coyote-wolf hybrids that are there now could be considered a sufficient partial replacement for the true wolf—a replacement just like all those *Canis* species down through time that followed the dog-faced cynodonts to keep predation pressure on large herbivores.

"We can't go home again." Evolution marches on, and we are part of it, a dominant part. There is no escape. We could, however, run one of the most interesting experiments in rebuilding adaptive gene pools ever undertaken. In the process, we might find out how to live better with wolves, coyotes, wild country—and ourselves.

LITERATURE CITED

Brown, J. H. 1995. *Macroecology*. University of Chicago Press.

Brundige, G. C. 1993. "Predation Ecology of the Eastern Coyote, *Canis latrans*, in the Adirondacks, New York." Unpublished Ph.D. thesis. College of Environmental Science and Forestry, State University of New York, Syracuse, New York.

Carson, R. 1965. *A Sense of Wonder*. Harper and Row.

Fuller, T. 1989. *Population Dynamics of Wolves in North-central Minnesota*. *Wildlife Monographs* 110.

Harrison, D. J., and T. Chapin. 1997. "An Assessment of Potential Habitat for Eastern Timber Wolves in the Northeastern United States and Connectivity with Occupied Habitat in Southeastern Canada." Wildlife Conservation Society, Working Paper 7. New York, New York.

Keith, L. B. 1983. "Population Dynamics of Wolves." Pp. 66–77 in *Wolves in Canada and Alaska*. Canadian Wildlife Service, Report Series 45. Edmonton, Alberta.

Mech, L. D. 1970. *The Wolf: The Ecology and Behavior of an Endangered Species*. Natural History Press, Garden City, New York.

Mlandenoff, D. J., and T. A. Sickley. 1998. "Assessing Potential Gray Wolf Restoration in the Northeastern United States: A Spatial Prediction of Favorable Habitat and Potential Population Levels." *Journal of Wildlife Management* 62:1–10.

Nowak, R. M. 1993. "Another Look at Wolf Taxonomy." Pp. 375–97 in *Ecology and Conservation of Wolves in a Changing World*, ed. L. Carbyn, S. H. Fritts, and D. R. Seip. Canadian Circumpolar Institute, Edmonton, Alberta.

Roy, M. S., E. Geffen, D. Smith, E. A. Ostrander, and R. K. Wayne. 1994. "Patterns of Differentiation and Hybridization in North American Wolf-like Canids. *Molecular Biological Evolution* 11:553–70.

Theberge, J. B., and Mary Theberge. 1998. *Wolf Country: Eleven Years Tracking Algonquin Wolves*. McClelland and Stewart, Toronto, Ontario.

Theberge, J. B., G. J. Forbes, I. K. Barker, and T. Bollinger. 1994. "Rabies in Wolves of the Great Lakes Region." *Journal of Wildlife Diseases* 30:563–66.

Theberge, J. B., and D. A. Gauthier. 1985. "Models of Wolf-Ungulate Relationships: When Is Wolf Control Justified?" *Wildlife Society Bulletin* 13:449–58.

Dreams of Wolves

Kristin DeBoer

I DREAM OF WOLVES

It is cold. Bone-chilling cold. I think the hair peeking from my hat is frosted. Snow squeaks with each step. I try to stay quiet because I am tracking. I have found a deer yard. Hoof prints are everywhere—so many it is hard to distinguish where one individual stops and another starts. Half-moons carved into the snow suggest a warmth only wild animals know in winter. I crouch down to pick up a scat. It is still warm. For a moment, fluttering streams of yellow light caress my face. I look up at the hemlocks. They are playing tricks with the sunlight, waving their feathery arms in the breeze.

    In that instant, I see them. One is peering from behind an oak. The other stands frozen with her head low, staring. I wonder where they came from. I wonder how long they have been there, tracking me. I wonder what to do. I stay on my knees and watch. They are taller than I thought they would be. Their thick gray and black fur radiates warmth. Their tongues hang out to the side, just like a dog's. I follow their white breath as it floats up to the sky. Their pointed ears twitch with each sound. I only can hear my heart pounding, my blood racing. Then they move their heads simultaneously, to get a better look at me. I wonder if they are sisters. I wonder who they are searching for. I wonder

how long they will stay. I wonder where they will go. There is only one thing I do
know. They are wolves.

Suddenly sound wrenches the forest. Maple and birch collide. The snow pack
shifts below me. The sun blackens. The wolves tear off. A buzzing gets louder.
With all the effort I can muster, I swat the alarm and lie back, awake. There
in my bed, I try to recall the wolves. I try to get back to the deeryard. The
images waver as if they were my memory of yesterday; then they are gone. The
dream seemed so real.

Despite a vivid dream life, this was the first time I dreamt of wolves. For
four years, I had been working as a wolf advocate. So it surprised me that
wolves had never graced my mind's nightscapes. But that month had been
an especially intense time for wolf advocates. The U.S. Fish and Wildlife
Service (USFWS) had recently given us some good news. They had prom-
ised to begin developing a wolf recovery plan for the Northeast in the winter of 1999. Although it was only January 14, the year already seemed to be
slipping by. Other wolf advocates and I hurriedly prepared to step up our
campaigns to ensure that the USFWS would take action. I had meetings
to organize. Reporters to contact. Activists to rally. Articles to write. Opponents to debate. Studies to commission. Brochures to design. With so
much to do, I didn't have time to linger over coffee and think about my
dream. I jumped out of bed, dressed, and went to the office.

It was not until much later that I took time to reflect on the dream. I am
not one to interpret dreams as symbols. To me, water means water. Sunlight is sunlight. Snow is snow. Wolves are wolves. The dream felt too real
to dismiss as a symbol for something else. It reflected reality, or at least what
reality could be. It could teach me something. It taught me that wolves
belong here. That dreams can be real.

The place I dreamt of was a section of the North Woods in southwestern New Hampshire, a place where I have spent time tracking animals and
exploring plant communities. Abundant populations of wolves used to live
there. But back then, several hundred years ago, it was a more expansive
and mature forest. Today, most trees are only fifty to one hundred years
old. Crumbling stone walls that crisscross the forest floor have left a declining mark of old farms and sheep pastures. Stump sprouts are a sure sign of
past logging.

By the 1800s, New England was about 70 percent deforested. Tall white
pines were cut for ship masts and timber frames. Maple and beech were

shaped into furniture in the factories of central Massachusetts. Farmers cleared whole hilltops and valleys to create fields where they grazed sheep. In *Reading the Forested Landscape: A Natural History of New England*, Tom Wessels points out that farmers occasionally spared individual trees to provide shade for the livestock. These open-grown trees, with their wide, sprawling branches, are sometimes referred to as "wolf trees," because like a wolf they stand alone.

As the land was cleared, wolves were cleared out too. One tale about New Hampshire's Mt. Monadnock offers a sense of the massacre. If you have ever climbed the mountain or seen it from a distance, you know that, although the peak stands at a relatively low elevation, Monadnock is bald. The story goes that before the turn of the nineteenth century a pack of wolves was still holding onto their territory, despite the townsfolks' best efforts. Livid about these "varmints" supposedly preying on their sheep, a group of people surrounded the wolves and herded them up the mountain. When they figured the whole pack was cornered, they lit the forested mountaintop on fire. Presumably, this pack of wolves burned in a fiery hell; the trees certainly did, and the bare granite crown stands as a reminder.

True or not, this story is an accurate account of the way the first European Americans embarked on a wolf extermination campaign. More hated than black flies and mosquitoes, wolves were poisoned, shot, trapped, burned, and scared away. Massachusetts Bay Colony was the first place in North America to place a bounty on wolves, in 1630. Colonists were awarded a typical month's salary in exchange for a few wolf heads. The rest of New England soon followed suit.

In 1864, writer and naturalist Henry David Thoreau traveled to the wilderness of northern Maine, where he experienced wolf country. He wrote in *The Maine Woods*, "It is a country full of evergreen trees, of mossy silver birches and watery maples . . . Such is the home of the wolf . . . What a place to live." The wolves he spoke of were probably some of the last packs in the Northeast. Soon thereafter, the species was entirely wiped out. The whole agonizing endeavor only took about 270 years—just a flash in time compared to the thousands of years wolves inhabited this region.

In New York State the last wolf was killed in 1897. I have seen this animal. It stands stuffed and sterile behind glass in The Adirondack Museum of Blue Hill, New York. When I first looked at the wolf I could almost imagine it alive. The fur still appeared soft, the teeth and claws sharp. The posture suggested movement. The ears looked alert. The art of taxidermy tried to trick me into believing the wolf was still there. But it was the eyes that gave the ruse away. They had been replaced by cheap glass imitations. Peering into that vacant stare, all I could see was a reflection of myself searching in vain for the real wolf. Along with those eyes, the species had lost its ability to see, to perceive the world in which it once lived. Blinded, the wolf's penetrating gaze has not been felt in the North Woods in over one hundred years.

I left that museum in horror. Imagine being the last of your kind. How long had that single wolf wandered alone without a family? Had she seen her pups murdered? How much time had she spent searching for another mate? How often did she howl only to hear the wind howl back? Without a pack to hunt with, had she gone hungry? Did she hear her killer coming? Could she know she was no longer the top predator? Did she understand this might be the end?

Unfortunately, the war on the wolf did not end here in the Northeast. Pioneers took this vengeance with them as they traveled West. In a quest to tame the wilderness from Massachusetts to California, they hunted down wolves in every forest, hillside, valley, peak, and plain. Wolf extermination was actually a full-time job. Professionals could destroy tens of thousands of animals per year with special traps and lethal poisons.

Halfway through the twentieth century, it was suspected that most of the United States of America was finally free of wolves. Only then, in the final hour, did we begin to awaken from the stupor of human superiority to realize the extent of our crimes against nature. One of our country's most renowned conservationists, Aldo Leopold, was among the first of those who sounded the alarm with his essay "Thinking Like a Mountain," published in *A Sand County Almanac* in 1949. But it was decades before something was done. The wolf was first protected in 1974 by the then new law, the Endangered Species Act. By grace, a few wolves were left to protect. Four hundred eastern timber wolves survived in the far northern forests of Minnesota—an area less than three percent of their former range in the lower forty-eight states. Somehow, they knew enough to hide.

DAYDREAMS OF RESTORATION

I work for the conservation group RESTORE: The North Woods. When RESTORE first launched our campaign to restore the wolf to northern

New England and New York in 1992, some critics called the idea unrealistic, far-fetched, crazy even. The more sympathetic just called it visionary. I prefer the latter. Occasionally, when the phones stop ringing at the office, I daydream about what the vision of ecological restoration really means. Sitting in front of my computer surrounded by piles of memos, stacks of documents, and a bulletin board covered with colorful images of the natural world, I gaze out the window and imagine.

Yearning to get out of the office and into the woods, I recall the times I have been in the presence of really big trees—not the astounding sequoias out west, but the remnant tracts of old growth in the east. Not many are left. Less than 1 percent of native eastern old-growth forests survive. Three-hundred-year-old oaks. Hundred-foot-tall white pines. Mossy old spruce. American beech so wide you need four sets of arms just to hug the trunk. Looking up into their high branches, I see wise old sentinels of wildness. I dream that these trees are not relics of the past but forests of the future. As we work toward this vision, these restored forests will be where wolves and other forest dwellers reclaim their home. They will be the forests where future generations of all species rediscover the wilderness soul of this land.

Yet, the voices of the forest remind me that many species have already returned. Common loon sounding a surreal laugh from a distant pond. The thunder of flapping wings as American turkey take flight into the canopy. The screech of a barred owl. Flocks of Canada geese honking overhead on a starlit flight. The forest's voices are calling out once again.

Subtle visual cues also signal that wildlife are coming home. The trail of a white-tailed deer winding among fallen leaves on the forest floor. Bonsai-like trees whose branches and crowns have been gnawed off by porcupine. A beech tree marked dozens of times by the claws of black bear returning to feast. Moose tracks sunk deep in the mud of a beaver swamp. Bald eagle soaring on a thermal, fading into the deep blue sky.

One hundred years ago, most of these species had nearly or entirely disappeared from the Northeast due to habitat destruction and overhunting. Today, with a little imagination, the forests of this region can seem so full of life that one can pretend we have traveled back in time, four hundred years ago to our native wild land.

The vision of ecological restoration is not only a dream of what could be; it is a process that has already begun. We do not only have to imagine. The seeds of a restored North Woods have sprouted right here in our backyard. Trees are growing back. Wildlife are returning. However, critical pieces are still missing, including the wolf. Although the process of natural restoration is tremendously encouraging, the rewilding of the North Woods in the northeastern United States will only be complete if we, the human inhabitants, realize our opportunity and responsibility to encourage nature to heal itself, to become whole once again.

During the last century, the recovery of wild nature has occurred only as a result of our choices—whether intentional or unintentional. Forests are coming back because early settlers abandoned old farms and fields. Large tracts of land remain undeveloped only because we choose not to develop them. The great Maine Woods, the Northeast Kingdom of Vermont, the North Country of New Hampshire have all been left relatively wild not because they are legally protected but because most people simply cannot afford to live that remotely. Benign neglect is a powerful ally of the natural world. We also owe gratitude to early conservationists and local citizens who led the way to protect some of the largest and most beautiful reserves in the region, such as the forever-wild Adirondack State Park in New York and Baxter State Park in Maine, and wilderness areas in the White Mountain National Forest of New Hampshire and the Green Mountain National Forest of Vermont.

Along with widespread reforestation, habitat is recovering too, providing shelter and food for wildlife. But in most cases, protective laws and active recovery programs have aided wildlife restoration. White-tailed deer, for example, nearly extinct here a century ago, are now common across the region because state wildlife agencies actively reintroduced the species. Beaver too have benefited from reintroduction programs. Once, this wetland-creating species was entirely wiped out by a lucrative international fur trade. Now population estimates have soared close to precolonial levels. Some species are more populous simply as the result of laws that safeguard them against overhunting and overtrapping. Moose, for example, once relegated to the wildlands of Maine, are steadily increasing in population at a rate of 10 to 15 percent per year because of a restricted hunting season. As a result, these gentle giants are migrating in search of new habitat, as far west as New York and as far south as Massachusetts.

Other species have not fared so well. Predators like the wolf, wolverine, cougar, and lynx have yet to be restored. Carnivores may have been left behind because they are often more sensitive to human disturbance, and so require more protection and secure habitat. However, carnivores are also discriminated against as a result of negative public attitudes. That these animals must kill for a living has made them less palatable. Usually, traditional fish and game policies favor the more docile herbivorous "game" species—Bambi and Bullwinkle. But perhaps the most significant reason the wolf and its cohorts have not returned is that their time has not been right, until now. Only because moose, deer, and beaver are abundant has the table been set for wolves to return. In fact, two separate scientific studies, one done by Harrison and Chapin and the other by Mladenoff et al., estimated that northern New England and New York have enough wild prey and open space to sustain over one thousand wolves.

The land seems ready to reweave one more strand in the revival of this region's biodiversity, to begin the dance between top predator and prey once again. The question is whether the human communities of this region are ready to welcome the wolf back home. For wolves and other top predators to recover, our cultural perception of New England must evolve with the land. The traditional view of this region is that of a pastoral landscape characterized by quaint towns with pretty white churches, town greens, and rambling country roads. Nature is important in this worldview, but it is a tamed nature, defined by lush farms and active woodlots. In this picture of domesticated bliss, it may be hard to imagine fitting in wolves and wilderness. But there is another side of New England—the wild side—where native wildlife and wild forests are making a comeback, trying to erase some of the indelible marks of human civilization. This is the New England from which a wilder future can unfold.

RESTORE, for one, has put forth a vision designed to inspire New Englanders to develop a cultural commitment to ecological restoration. The introduction to our brochure reads:

> We are five centuries too late to *save* the primeval North Woods. But we still have a chance to *restore* it. Once a great wilderness, stretching from the Atlantic Ocean to the Great Plains. Then, a region seriously damaged by industrial development. Today, a land where wild nature is struggling to return. Tomorrow, America's first restored landscape; a place of vast recovered wilderness, of forests where the wolf and caribou roam free, of clear waters alive with salmon and trout, of people once again living in harmony with nature.

In the last decade, this vision has started to take root, and it is transforming the New England conservation agenda. In the 1990s, several other visionary groups with similar missions were established, including Conservation Action Project, Forest Ecology Network, Forest Watch, The Greater Laurentian Wildlands Project, Keeping Track, Northern Appalachian Restoration Project, and Wild Earth. Together, we are creating a grassroots movement comprising thousands of citizens who want to live in a wilder New England where vibrant towns and small farms are infused with and surrounded by wildness.

It is this broader mission to craft a wilder future that has helped the specific goal of wolf recovery to flourish. Wolf recovery is not just about restoring the wolf. It is about beginning to reweave the whole fabric of life. It is about biodiversity and wilderness restoration. It is about how we define our role in nature. It is about our worldviews. It is about lifestyles. It is about values.

Michael Kellett, the founder and Executive Director of RESTORE, who has a talent for coming up with revolutionary ideas, has said, "RESTORE chose wolf recovery as our first campaign because we realized that the idea of wolves in New England could help us ignite a wilderness restoration movement. The wolf's ecological role is essential. But just as important is the wolf's symbolic connection to wilderness. If wolf recovery is done right, it will not only benefit the health of our ecosystem, it will help change our relationship to the land."

Nowhere in America is there a better opportunity to restore an entire landscape to ecological health than in the great North Woods. Author Bill McKibben has written that the large-scale reforestation of the eastern United States is the most important environmental story in the nation. And the story is only beginning. The essence of the primeval North Woods landscape endures in scattered old-growth groves, vast second-growth forests, and undeveloped lakes and rivers. Wolves, lynx, cougar, and salmon still have a chance to repopulate their old haunts. We can restore much of the region to its former natural grandeur, but we need to act soon, before the opportunity is lost.

The people of New England and New York have a strong conservation ethic. Yet, with a few exceptions, conservationists have focused largely on human-centered environmental issues, such as toxic pollution, organic farming, environmental justice, public transportation, sprawl, and sustainable logging. While each of these conservation efforts is essential and complementary to ecological restoration, Northeasterners have largely consigned national parks, wilderness, and endangered wildlife protection to far-off places such as the West, Alaska, or even South America. As a result, large-scale ecological restoration is often considered politically unrealistic in this part of the country.

But political reality can and must conform to ecological reality. Through RESTORE's work, we are showing the people of the Northeast that it is possible to restore our own landscape right here at home. We are offering positive programs for creating new parks and wilderness and bringing back the full diversity of native wildlife species, starting with the wolf. This is a very long-term effort. But with a little luck and a lot of vision, there is hope that the North Woods will be wilder at the end of the twenty-first century than it is at the end of the twentieth century.

At the beginning of a new millennium, the mainstream New England conservation agenda is starting to incorporate these ideas. For example, the State of Maine under great pressure from conservation groups, including Atlantic Salmon Federation, Conservation Action Project, Defenders of Wildlife, RESTORE, and Trout Unlimited, has begun the long process of restoring dwindling populations of wild Atlantic salmon. Well-established private land trusts like The Nature Conservancy have succeeded in raising millions of dollars to protect from development hundreds of thousands of acres of the Maine Woods. Most of the major conservation groups in New England, as part of the Northern Forest Alliance, have embraced the need to protect a system of wildland reserves across the region.

The idea of restoring the wolf has helped to reinvigorate our conservation consciousness, to broaden our goals to include the restoration of all native species. While wolf recovery alone cannot restore a degraded, fragmented ecosystem, it is one way to start restoring wildness to our forests. And as the wild seeps back into the forest, it will start to seep back into our souls. That it is what it will ultimately take—courage born of our deeply felt convictions—to complete the job of restoring the wild side of New England.

Stephanie Mills, in her book, *In Service of the Wild: Restoring and Rehabiting Damaged Land*, leads readers to a deeper understanding of how we can develop a personal commitment to the work of restoration. She writes:

> Remember . . . how good it is to wander in the woods, or the meadow, or down by the river; how it feels to be baking in the desert or sinking in the swamp or creeping through the forest or standing with the wind in your face in the big middle of the plains. . . . Remember what a relief it is, what a joy, to

find yourself sentient in the company of wild things. . . . And if you can't remember, imagine it. Imagine the night being not silent nor humming with
traffic, but rich with sounds of birds and frogs and insects, with stirrings of
nocturnal mammals, all manner of vitality. Imagine a world where the life of
the Earth and of the human spirit could go on, evolving, diversifying, adapting, changing, and surprising, fearlessly; if it can be imagined, it can come to
be. If it can be recalled, it may be restored.

Time and time again, wildlife biologists have proven it is technically possible to successfully restore the wolf to its former habitats across the country.
However, wolf restoration requires much more than scientific expertise. It
requires vision and imagination. Welcoming the wolf back home invites us
to dream a new future—a wilder New England where wolves and other
wild creatures roam free.

DREAMING OF WOLVES IN WILDERNESS

*The first time I experienced wolf country, outside of my dream world, was in
Isle Royale National Park in the North Woods of Michigan. My husband,
Tom Hartman, and I took the long journey there across Lake Superior by
ferry. The day we traveled, the lake was as smooth as satin. As we sat on the top
deck, a cool, still fog surrounded us like a cloud of silence. The only sound was
the bow cutting through the water. Moist air veiled my eyes. I could only make
out filtered colors—the green boots of the person sitting beside me, Tom's red
hair, the yellow jacket of a passenger by the rail. We traveled in that surreal
world for five hours. While Tom chatted with one of the crew members, I sat
serenely staring into the milky water and dense fog. Slowly, I began to shed lay-
ers of urban life and work. By the time we arrived on the island my thoughts
had drifted far from my domestic routine, as we entered a dreamy land called
Wilderness.*

*As we donned our backpacks, it became clear that the glassy lake was really
the calm before the storm. From that moment on and for three days more, tor-
rential rains soaked us day and night. If the wolves who lived here had any
sense, they were under cover, but we trekked along in search of them anyway.
Hiking along the coast of the island, the lake crashed onto the shore like a rag-
ing sea. Sheeting rain pelted our heads and seeped under our collars. Deep
puddles overflowed into our boots. Spruce trees shed water like streams of tears.
White birch wailed in the wind.*

On the fourth morning, our wet bodies, packs, and tent were finally warmed with rays of sunlight. At once, this "sweet wild world," as Thoreau so aptly described it, was transformed. Butterflies came out of hiding to dance in the meadow. Wood thrush broke the silence with long musical trills, as wind whistled along through the wet treetops.

After we set up camp, we wandered down to an inland pond to fetch some fresh water. The rain had produced lots of mud, ideal for tracking. Yet the only fresh tracks we spotted were our own. Without a lead, we returned to camp for dinner. Perhaps wildlife would come out later in the summer dusk—after ten o'clock this far north. A few hours passed while we soaked up the sun. Then we walked down to the pond again.

There on the trail, we spotted our first sign of eastern timber wolf—a wolf print almost 5" long—right on top of one of our own boot tracks! Just a few feet ahead, we found more prints, moose this time, a mother cow and calf. Excitement whirled through my body. Had we stumbled upon an act of predation? Had this wolf practically walked through our camp in pursuit of its prey? Had he been watching us when we walked to the pond just hours before?

We continued cautiously along the path, eager to see the animals, but not wanting to scare them, or ourselves. At the crest of the hill, the pond came into sight. There we watched the moose cow and calf grazing knee-deep in the water —taking only quick moments to dip their heads for some browse. They seemed intensely alert, much more aware than we were. Then a twig cracked in the nearby underbrush, as if someone or something had stepped on a branch. Our heads jerked toward the sound. The moose plunged into the water and swiftly swam across the pond—probably to escape the wolf. We were left in awe and sat on a rock gaping. Was the wolf still watching us now?

The wolves of Isle Royale live in a Wilderness with a capital W. Most of this national park is protected by the 1964 Wilderness Act. Under this law, Wilderness is defined as "an area where the earth and its community of life are untrammeled by man, where man himself is a visitor who does not remain." In the United States, about 104 million acres are federally protected Wilderness, and the National Park System protects approximately 83 million additional acres. Together, about 187 million acres are off limits to most forms of resource extraction, such as logging, mining, and development. Our nation's wilderness and national park systems are a great accomplishment to celebrate. Yet, less than 10 percent of our once wild country is protected; millions of acres more could and should be preserved and restored.

Like the wolf, wilderness is endangered. The wolf wars were just one part of a grand scheme to tame the wild. In this modern, technological, and commercial world, wildness is vanishing because humans have tried to take control. On an earth that is dominated by us, we see farmers as the producers of food, loggers as the keepers of the forests, hunters as managers of wildlife. We too are creatures that must meet our own needs for food, shelter, and water. But when we try to control nearly every natural process, the wild is transformed to meet our own demands, not the needs of the other creatures with which we share the land. Human-dominated landscapes depend upon people to maintain them. Even with all we know about nature, it is impossible for humans to mimic the complexity of the natural world. As a result, wild systems are homogenized and biodiversity is often simplified or lost.

Wilderness is a place where humans yield control. Whether in our souls or on the land, it is where we invite nature to take over. In the wild, food grows by itself, forests succeed on their own, predators and prey manage each other, natural processes prevail. Without us there to control the fields, tree farms, and game, they will "go wild." Wilderness exists because humans do not try to manage it, manipulate it, exploit it, or control it. Wilderness exists whether or not we do. It precedes us. It will follow us. As the ecologist and theologian Thomas Berry writes in *The Dream of the Earth*, "the earth is primary; humans are derivative." Wilderness is the raw earth from which the splendid diversity of life has evolved. We need wilderness for the diversity of life to continue.

Wilderness needs wolves. The presence of wolves enables the wilderness of the Northern Hemisphere to function as it has for millennia. In the wilderness of Isle Royale, wolf biologist Rolf Peterson has conducted one of the longest-running studies of the predator-prey relationship between wolf and moose. He has shown that wolves help maintain the health of the forest by controlling the population of their prey. Before we had set out on our trip to Isle Royale, the park ranger explained how the moose population crashed during the winter of 1995–96, plummeting from about 2,400 animals to a mere 500 in one season. A catastrophic crash of this magnitude had not been documented since 1934 when the moose had no natural predator. Wolves did not migrate to the island until the 1940s.

Scientists believe that the moose became overpopulated in the early and mid-nineties because the wolf population had been drastically reduced a decade before. Apparently, the wolves had been dying off from parvovirus

disease and inbreeding. Without enough wolves to consume them, the moose set out to consume the forest. These large ungulates eat between 35 and 50 pounds of vegetation per day. On an island only about 45 miles long and 9 miles wide, it did not take long for them to overexploit their food supply. The forest suffered and so did the moose. As the long, hard winter complicated their survival, most of the moose starved to death. We saw those moose—dozens of skeletons littered throughout the forest floor, jawbones catching puddles in the rain.

A wilderness without wolves would be incomplete, and so would wolves without wilderness. Across the northern hemisphere, the wolf thrived for eons on a wild earth. Over time, dozens of subspecies of the wolf evolved to be perfectly adapted to a distinct habitat and diversity of creatures. Wilderness was their home. Wilderness was their refuge. As wolves were exterminated across this country, the only places they survived were the remote wildlands of Minnesota and Canada—places where they could not be reached.

Today, wilderness areas, national parks, and other protected reserves still provide the best habitat for wolves because they offer security from human threats. Cars and guns are two of the highest causes of wolf mortality. Among the wolves reintroduced to Yellowstone National Park, two-thirds of all human-caused mortalities have occurred outside the national park borders. Since national parks and wilderness areas restrict vehicular access and hunting, the risk of a wolf getting killed accidentally or intentionally is reduced. A pack of wolves requires up to 100 square miles of territory, depending on the prey base, so the bigger those restricted areas are, the better. One reason wolf recovery programs have been so successful in and around large areas of core habitat such as Glacier National Park, Boundary Waters Wilderness Area, and Yellowstone National Park is that they help assure the wolves' safety.

In the Northeast, there is probably enough remote habitat for wolves—right now. However, most of that habitat is not guaranteed for the future. In this corner of the country, we have only one small national park and few wilderness areas. Second-home development and large-scale industrial logging are rapidly eroding the wildlands that are left. Many conservationists believe that habitat protection is one of the fundamental challenges for wolf recovery in this region. So several conservation groups working for the wolf are also working on complementary efforts to protect land.

One of the boldest land protection proposals within potential wolf reintroduction areas is the creation of a new 3.2-million-acre Maine Woods National Park and Preserve. Bigger than Yellowstone and Yosemite combined, this national park would provide a unique and exciting opportunity to restore wilderness on an Alaskan scale. As the "Yellowstone of the East," this national park could become a model of ecosystem recovery, providing a sanctuary for all the native wildlife that once called this region home, from wolf, to moose, to woodland caribou, to salmon, to loon. Here individual wolves would be afforded the quality habitat they deserve. And with a protected contiguous core as their base, the population as a whole may have a greater chance to become permanent residents.

Establishing the proposed Maine Woods National Park is not a prerequisite for wolf recovery to occur. However, over the long term it can only help to ensure the viability of the species and provide higher quality habitat. Presently, the Maine Woods, as the largest remaining tract of undeveloped forest east of the Mississippi, is the most suitable habitat for wolves in this region. Most of that land, over ten million acres, is privately owned by a handful of multinational timber and paper corporations. Accordingly, very few people permanently reside here and it remains a remote place where wolves could have enough space and sufficient prey to survive. But this ownership structure also gives the forest products industry virtually complete control over the current and future status of wolf habitat. Under their tenure, an area larger than the size of Delaware has been clearcut in the last fifteen years; thousands of miles of logging roads have been built, enough to wrap around the equator; vast monocultures of pulpwood have been planted; and herbicides are periodically sprayed to suppress the growth of "undesirable" trees.

Although wolves could exist here, this degraded forest does pose some challenges for recovery. Logging roads provide access for people who would rather not see wolves restored. Clearcutting has impacted one of the wolf's primary prey species by reducing over-wintering habitat for white-tailed deer to just 2 percent of the forest. What is worse is that the Maine Woods is increasingly at risk of wholesale development and fragmentation. From June 1998 to March 2000, about 3.8 million acres, or 18 percent of the State of Maine, have been sold off to the highest bidder. Each large land liquidation puts wolf habitat into jeopardy as the risk of development looms closer.

This upheaval in land ownership is unlike anything seen in Maine for two centuries or more. However, the crisis also offers an historic opportunity. These lands could be acquired at fair market value from willing sellers for only $250 an acre. For less than the cost of one B-2 bomber, about $1 billion, the public could purchase and protect the heart of this region as a new national park.

The Maine Woods National Park is still a dream, and it will need the active support and enthusiasm of the American people to become reality. But local and national support are growing in leaps and bounds. Originally proposed in 1994 by RESTORE: The North Woods, now the national park idea is endorsed by scores of conservation groups, hundreds of businesses, and dozens of prominent scientists, celebrities, conservationists, writers, and political leaders.

As grand as it sounds, though, even a 3-million-acre national park alone may not be enough for wolves and other wildlife to live out their lives safely and freely throughout the region. This is only the beginning of what wilderness advocates envision for the Northeast. The Maine Woods National Park could become the foundation of an even larger system of interconnected parks and reserves stretching across the North Woods from Maine to the Adirondacks and across the border to the wildlands of Canada.

In fact, recent studies have shown that wildland connectivity is essential for wolves. Even the Northeast's largest existing wilderness area, Adirondack State Park, may not provide good enough habitat for wolf restoration unless more extensive conservation planning is put into place. A 1999 study conducted by the Conservation Biology Institute and sponsored by Defenders of Wildlife on the ecological feasibility of wolf reintroduction into Adirondack State Park concluded that:

> Though our analyses suggest that the AP [Adirondack Park] comprises sufficient habitat to support a small population of gray wolves, regional conditions are not conducive to sustaining wolves over the long-term (e.g. 100 years). Given current trends in regional development, we anticipate environmental conditions necessary to maintain wolves will deteriorate over the next 100 years.

This study does not rule out the potential for wolf reintroduction into the Adirondacks. But it does raise the bar on conservation efforts if wolf recovery is truly to succeed. Protecting isolated wilderness preserves is not enough. Actions will need to be taken to maintain the integrity and security of wolf habitat across the ecosystem within and outside public reserves. New national parks and expanded wilderness areas, as well as other connected wildland reserves, will be essential to protect large areas of contiguous core habitat where wolves and other species can thrive now and far into the future.

Planning for ecosystem integrity based on the needs of wide-ranging carnivores like the wolf is the work for which The Wildlands Project (TWP), a North American conservation organization, is best known. Co-founded by the visionary and stalwart activist Dave Foreman, and the father of conservation biology, Dr. Michael Soule, TWP has inspired conservation groups across the continent to think bigger and take bolder action on behalf of biodiversity. TWP's Northeast regional effort, the Greater Laurentian Wildlands Project (GLWP), is collaborating with scientists and concerned citizens to design a system of core wilderness reserves connected by buffers and corridors across New England, New York, and southeastern Canada. According to Paula MacKay, GLWP's point person for wolf recovery, "if wolves are to be successfully restored to their vital role as a keystone predator in the Northeast, we are morally obliged to provide the secure habitat necessary for their long-term survival. We need more parks and wilderness areas to protect wolves and other sensitive species, but we must also ensure that such ecological refugia aren't islands in a hostile sea.

By restoring habitat linkages between core reserves and working to foster good stewardship and human tolerance across the landscape, we can create a web of ecological jewels that will accommodate wolves and the inter-dependency of all living things."

Some critics say that The Wildlands Project vision is too broad and long-term to be applicable in the here and now. Yet, thinking on an eco-logical scale may come more naturally to mere humans than it appears at first glance. The Vermont-based nonprofit organization Keeping Track, for example, is training hundreds of local citizens throughout the North-east and portions of the West to detect and record evidence of large and wide-ranging carnivores, enabling volunteers to become involved in the appropriate long-term stewardship of wildlife habitat. Data are collected seasonally, year-to-year, and are increasingly important to local and re-gional planners and wildlife managing agencies.

Susan Morse, founder and program director of Keeping Track, is un-abashedly committed to protecting biodiversity. She believes that grass-roots citizen-based fieldwork and resulting conservation advocacy are critical components of what she calls "the vital work to be done." "We must find ways for more of us to participate in the planning process and draw the line on so-called development as well as unacceptable and unsustainable exploitation of natural resources, or all life as we know it faces a miserable future. Considering the needs of the wolf, lynx, bear and other carnivores will appropriately guide us as we seek to understand ecological economies, make necessary sacrifices, and learn to think on a grand scale."

Long-term visionary projects like the Maine Woods National Park and those sponsored by The Wildlands Project and Keeping Track unite the idea of wolf recovery and wilderness restoration in a synergy that inspires people. Perhaps it is because they tap into a deep human desire to feel more connected, to understand how the lives of wild animals weave together with our own lives and those of our children and our children's children. Or perhaps it is because wolves and wilderness are inextricably bound in time, in place, and in the human psyche. Just as wolves and wilderness were tamed and dominated in parallel, both forces of life can be truly revived only if they are restored together. Wilderness is where wolves can most fully be wolves. Without wolves, wilderness is not truly wild.

However, some conservation strategists believe that wolf recovery and wilderness protection should not be connected. Unlike most endangered species recovery efforts, unfortunately the work of wolf recovery has drifted away from habitat protection in recent years. This trend is, in part, due to the recent expansion of wolf territory into domesticated areas of Spain, India, and Italy. The experience in these countries has proven that wolves are generalists that can survive in a range of habitat types, not specialists that depend on pristine wilderness for their very survival. The numbers of wolves have increased in areas where there is logging, hunting, agriculture, and even development, as long as there is sufficient prey and social tolerance among the human residents. But more likely, the shift away from habitat protection has even more to do with the tidal wave of political pressures moving against endangered species and wilderness. In the end, the divorce of the age-old union between wolves and wilderness may have more to do with political reality than ecological reality.

Although it may be an acceptable beginning to restore wolves to less than wild land, ultimately wolf recovery must also honor the wilderness heritage of the species. Otherwise, wolves and ecosystems will suffer. If the goal is simply to get wolves back in the short term, then protecting wilderness habitat may not be necessary. But what good is wolf recovery if we do not simultaneously work to restore the ecological integrity of the North Woods? Indeed, the point of wolf recovery is not only to ensure the survival of one species, it is to begin restoring biodiversity—whole and healthy ecosystems complete with all of their wonderful parts. As Douglas Chadwick writes in his book *The Kingdom*, "A fundamental issue revolves around wolves. Are we really out to conserve wildness? Or only the pieces that suit us? To the extent that we pick and choose pieces, wildlife communities will reveal that much more about the temporary goals of human communities and that much less about the holistic workings of nature. How, then, shall we ever understand creation?"

Wolf advocates cannot afford to take the path of least resistance. We owe it to the wolves to ask the harder questions, to do the harder work, to do what will be best for wolves and the full range of native wildlife for generations to come. If we dream of restoring the wolves to a vast recovering wilderness, and act on it, then we will have done right not only by the wolf but by the lynx, cougar, elk, moose, bear, wood thrush, salmon, deer, beaver, grouse, and ourselves. The wolf can help usher wild nature back as it returns to this corner of the world, but the rest of it will be up to us.

WOLVES BEYOND WILDERNESS—
BAD DREAM OR REALITY?

In a cool office lit by low fluorescent lights, I watched over the shoulder of a wolf biologist while he circled areas on a map. He seemed to be a mild mannered, rather mellow guy, but his eyes lit up when he talked about red wolves. He explained that the U.S. Fish and Wildlife Service knows where the wolves are because many of them have radio collars. The animals' movements are monitored 365 days a year to measure their recovery progress. When he was done with his orientation, he offered me a few directions and sent me on my way.

It was extremely hot and humid that August day—though I suspected it was always this way in North Carolina—so I indulged in air conditioning while I drove. From behind the bug-stained window of the rental car, I crossed over a major road or two, passed a diner, then a trailer park, and navigated through the pine plantations before arriving at my destination—fields and fields of agricultural crops. This was wolf habitat? It certainly was not wilderness. It wasn't even a tree farm!

In an effort to recover from the shock, I got out of the car and began searching for wolf tracks. Secretly, I hoped I wouldn't find any. I wished that I had found the wrong place. I prayed aloud, "Please don't tell me this is where the wolves live. I drove here, for god's sake, in an air conditioned car!" It was just too bizarre. I trudged along slowly because of the heat, searching along the dirt road that sliced through the middle of geometric rows of crops. I tried to identify what kind of plants these were, but they seemed foreign to me. Cynically, I thought they were probably some genetically altered hybrid born in a laboratory at Monsanto.

Then I noticed a track, the unmistakable print of a large canine. At first I thought it was left by a dog. But as I followed the trail, I realized that the animal's path was straight and direct, a defining characteristic that distinguishes determined wolves from meandering dogs. Now there was little doubt in my mind this was a red wolf. I looked up to take note of how far I had walked only to be confronted with a bright orange warning flag. Danger. With dismay, I wondered if the area had been sprayed with pesticides.

Unfortunately, this was not a bad dream. It was reality. This was red wolf country in North Carolina. By 1980, the red wolf, *Canis rufus*, was declared extinct in the wild. Like gray wolves, red wolves were persecuted through overhunting and excessive trapping and much of their habitat was destroyed. To save the species, fourteen red wolves were rounded up from the wild in Louisiana and Texas for a captive-breeding program. From 1987 to 1994, sixty-three captive-born red wolves were released into parts of their former habitat—over 330,000 acres of forests, swamps, and fields in the Alligator River and Pocosin National Wildlife Refuges, as well as adjacent land owned by the Department of Defense and private land owners.

The red wolf program has demonstrated that wolves, at least in the short term, technically do not require wilderness to survive. In fact, in a USFWS description of the red wolf recovery program, the agency points out that:

> Red wolves can flourish in a wide variety of habitats. Hence, insufficient habitat and habitat destruction per se are not limiting red wolf recovery. There is sufficient habitat available to meet the population objectives outlined in the Recovery Plan. However, much of that habitat is privately owned. Landowner support is therefore required for successful reintroduction initiatives. Thus, unlike most endangered species, recovery of the red wolf is not so much dependent on the setting aside of undisturbed habitat as it is on overcoming the political and logistical obstacles to human coexistence with wild wolves. This may also be true for reintroductions of gray wolves in some areas.

Under these circumstances, fighting to return from the brink of extinction in just a remnant of their former habitat, red wolves are adapting as well as can be expected. As of 1994, about thirty-seven pups were born into the wild. Today, according to the USFWS about sixty to one hundred red wolves survive in all. Although this is just a minute fraction of their historic population size, the good news is that these wolves have been saved from oblivion and released from the zoos into a relatively wild landscape.

Like the red wolf recovery zone, much of the available wolf habitat in the Northeast is not a pristine wilderness. And while not nearly as domesticated as the red wolf habitat I witnessed, the industrial working forest in the Maine Woods, in particular, has been dubbed "the paper plantation." In the eyes of the paper industry, the fact that wolves can survive on their used and abused land is "Mother Nature's seal of approval" for their industrial forestry practices. If the paper companies allow wolves to return, they will probably use that line to boost their environmental image. Yet, just because wolves can survive in their clearcuts does not mean that is the best possible habitat for them. After all, humans can survive in slums with atrocious housing conditions, yet no sane person would ever willingly choose this as his or her home.

The ability of wildlife to live outside wilderness, in places like tree farms and town forests, is perplexing. On the one hand, it is encouraging to see wildlife populations expand, because it may indicate that some land is becoming wilder. But in other cases, the reason wildlife endure is that certain creatures are able to cope by making do with what they have, just to hang on. The knowledge that some wildlife can live amongst us may cause our society to breathe a collective sigh of relief. It lets us off the hook, in a way. It allows us to indulge, without thinking as much about the impact of our actions. It is easy to get excited by spotting a red-tailed hawk by the side of a highway or seeing huge flocks of Canada geese foraging on a golf course. But is this really the picture of a conservation success story? These creatures have little choice.

Wildlife may be able to *survive* in human-dominated landscapes, but that does not necessarily mean that they can *thrive*. They can survive only if they are able to adapt to the limits we have set for them. The presence of people and all that we come with—fast cars, ATVs, SUVs, loud snowmobiles, guns, sprawl, pollution—can profoundly affect the behavior, social structure, and genetic variation of wildlife. In essence, wild animals can become different creatures in response to human pressures. Just as people's personalities, behaviors, and tastes can transform depending on whether one lives in a big city or small rural town, wild animals may also change their behavior whether they live in a tree farm or a large national park. The bottom line is that wolves need places where they can act like wild wolves. And only protected wilderness areas can provide that.

But when animals willingly choose to migrate beyond wild areas, beyond our human-imposed boundaries, to more human-dominated landscapes, who are we to say they should not? In October 1999 human population reached 6 billion people. In the next 30 to 50 years, that number is expected to almost double. As a consequence of population explosion and conspicuous consumption of natural resources, we may lose to extinction another 30 to 50 percent of the species on Earth. As human population growth soars, the reality is that we must learn to coexist with the remaining species across the landscape if we want to give wildlife a fighting chance.

Wolves themselves are spreading far beyond wilderness areas in northern Minnesota, Wisconsin, and Michigan to inhabit open space quite near suburbs. In Montana, a pack of wolves has wandered out of the Glacier

National Park to take up residence on a nearby ranch. Each year as wolf populations become more established in Yellowstone National Park, more and more animals are migrating outside the park boundaries to the north, south, east, and west. Maybe it does not serve wolves or humans to impose strict black-and-white divisions between where wolves should and should not live, between where wilderness begins and where it ends.

However, in wolf recovery programs around the country, decisions are routinely made that restrict where individual animals can and cannot go. When wolves move beyond predefined recovery areas and dare to trespass on private land or in more developed areas, they are often relocated, or worse yet, destroyed. In the red wolf recovery program, for example, the USFWS reintroduced captive wolves into North Carolina. But an adjacent state, Virginia, was not a participating party. Apparently the wolves were not told about this part of the arrangement, because one wolf wandered right into Virginia. The trip probably only took a day or two. Once the wolf was detected, the Virginian authorities called up the USFWS and had the illegal alien deported back to whence it came. Outrageous as it sounds, this was not an isolated case.

In the summer of 1999, a nearly identical incident occurred in the Yellowstone–Central Idaho wolf reintroduction program. This wolf migrated out of the "recovery zone" into eastern Oregon. Here the authorities were nervous about the possibility of the wolf attacking the plentiful cow population in the region. The animal was retrieved and sent right back—despite the fact that the wolf had not harmed any domestic animals.

If wolves are reintroduced into the vast wildlands of the Maine Woods and the Adirondacks, some will also surely roam into the smaller, less wild areas of potential habitat in the Northeast Kingdom of Vermont and the North Country of New Hampshire. They may even migrate south to the more populated parts of Vermont, New Hampshire, and perhaps Massachusetts. If this happens, will Massachusetts be the next state to demand the removal of an uninvited and unwanted guest?

While it is certainly the best starting point to restore wolves to the wildest lands, we cannot limit our vision of wolf recovery to the patrolled borders of wilderness areas. If we do, we may end up with little more than open-air zoos where wolf populations are intensively managed, manipulated, and controlled. To craft a successful wolf recovery program in the Northeast, we must consider how wolves and humans can coinhabit all sorts of land, from public to private. Wolves, like other wildlife species, must decide for themselves where they can survive. We need to imagine where wolves might roam in the best- and worst-case scenarios—so that we can help them become successfully re-established in a continuum of habitats, within wilderness and beyond.

One organization that is leading the way to promote a more enlightened view of wolf recovery is The Predator Conservation Alliance out of Bozeman, Montana. In the spring of 1999 they published a critical assessment of the wolf reintroduction program in Yellowstone National Park and central Idaho, entitled "At a Crossroads: The Wolf and Its Place in the Northern Rockies." In it they criticize the USFWS for restricting wolves to a "few token animals in isolated protected areas." As a solution they advocate for a recovery program where "wild and free-ranging populations [are] free to determine their natural patterns of abundance and distribution across the northern Rockies." Specifically, they recommend that "wolves should be allowed to live in all areas of suitable habitat where they can avoid conflicts with people and livestock. This means all areas of undeveloped lands, including private lands where landowners will tolerate them." But it is not up to the wolves alone to figure out how to avoid conflicts with human activities. It is up to people to modify some of our behaviors to prevent, reduce, and resolve conflicts as well.

In the Northeast, one place where it will be critical to minimize potential conflict is on the farms scattered within potential wolf habitat areas. To accomplish this, Defenders of Wildlife, a conservation organization nationally known for its wolf recovery initiatives, has sponsored a program to compensate farmers for livestock losses caused by wolf depredation. Fortunately, this is an affordable program, since wolves rarely prey upon domestic animals such as cows, turkeys, sheep, pigs, and dogs. Less than 1 percent of all losses of domestic animals is caused by wolves. However, the impact on one family of losing a well-cared-for and loved animal is significant. Compensation at a fair market value can at least make up for the financial loss. Funds are also awarded to livestock farmers who want to invest in preventive methods such as guard dogs and special fencing. Nina Fascione, the director of Northeast wolf recovery efforts for Defenders, is fond of pragmatic solutions like these. She, along with her organization, hopes that this program will provide the insurance that farmers need to feel comfortable with the possibility of wolves in or near their backyard. As Nina has said, "We believe wolf recovery can be successful without forcing people to negatively change their lifestyles, without having to fence off wilderness areas."

Agricultural interests will certainly play an important role in Northeast wolf recovery. But the majority of potential wolf habitat is actually on private timber lands. Since these family and corporate land owners have a good deal of political influence, regulations that force private land owners to change the way they practice forest management are seen by some as tantamount to burning the American flag. So none of the wolf advocacy groups in the Northeast advocates for using the wolf to regulate private land. Instead, some conservation groups, such as the National Wildlife Federation (NWF), are taking the lead to encourage private landowners to *voluntarily* allow endangered wildlife, like wolves, to take up residence on their land. One way NWF is approaching this challenge is by promoting a forest certification called SmartWood that recognizes land owners for using sustainable forestry methods that also sustain wildlife habitat. Certification requires that "management addresses the conservation of threatened, rare, endangered and unusual plant and animal species, natural communities and critical habitats." Green certifications like these can give businesses a competitive advantage in a market where consumers are increasingly interested in where their wood comes from. And it may give land owners the incentive they need to cooperate in wildlife recovery efforts. Liz Soper, the wolf coordinator at NWF, has said that her organization is showing land owners that "sustainable forestry and wolves can go hand in hand."

Similar motivations have prompted other conservation groups to work with private land owners to purchase conservation easements on their land to secure it from being developed. Although logging is usually still permitted under this arrangement, maintaining the land as "working forest," rather than subdivisions or strip malls, is meant to help maintain a less fragmented and degraded habitat for wolves and other wildlife.

The USFWS has also begun to devise ways to promote wolf recovery on private lands. They are proposing to downlist the protected status of the eastern wolf from "endangered" to "threatened." Certainly, no scientific or legal basis supports this proposed action. No self-sustaining wolf populations are left in the Northeast to justify this downgrade in protection status. Instead, the proposed downgrading is a political move designed to lessen opposition to a wolf recovery program. Although "threatened" status theoretically brings nearly the same level of protection for wolves, it also allows for more flexibility within the Endangered Species Act. Flexibility, in this case, would give the state wildlife agencies and corporate land owners more control over if and how wolf recovery proceeds.

The USFWS insists that this proposed concession is necessary to move the effort forward by helping private land owners to understand that wolf recovery will not force them to change their land management practices. But wolf advocacy groups remain concerned that this approach might mean extensive manipulation and control of recovering wolves. A certain level of pragmatism is necessary to achieve wolf recovery over the long run. However, there is little evidence that this proposed downlisting is necessary. Indeed, it could well undermine the chances of developing an ecologically sound recovery program where wolves are given the right to roam wild and free.

Efforts to allow wolves to live in a human-dominated landscape represent a compromise between culture and nature. Such compromises may help wolves and humans to peacefully coexist in today's real world. They begin the course toward making agriculture more predator-friendly, toward making logged forests more acceptable, toward making wildlife more diverse. And, coupled with initiatives to protect and restore vast wilderness areas, they will help us navigate the long, steep, and winding trail toward fulfillment of the dream of ecological restoration.

DREAMS OF A NEW LAND

The wolf populations nearest to the northeastern United States live only seventy miles north of the border in southeastern Canada. Last summer I traveled there to understand what challenges wolves face in Canada and to witness for myself the obstacles that might be preventing these wolves from migrating south in search of new land.

My destination was Algonquin Provincial Park in Ontario, which harbors the nearest wolf population to the proposed wolf reintroduction area in Adirondack State Park of New York. There, I met up with a friend and colleague, Kathleen Fitzgerald. Kathleen, a passionate activist who has made wilderness restoration her life's work, was just finishing up the Field Naturalist graduate program at the University of Vermont. As part of her training, she had been doing wolf research here in Algonquin and other parts of Ontario and Quebec for the past four months. With her newfound field expertise in Canada, she made a perfect guide for an excursion into this wolf country. So, with two other friends, Kathleen and I set out on a five-day journey by canoe.

Several hours into the trip, the rhythmic action of paddling nearly put me into a trance. Deep indigo skies, bright green hills, and silvery sparkling water were etched into my mind so clearly that, later that night, when I closed my eyes, I still felt as if I was floating in a sea of blue, green, and silver.

After our first day's paddle, we all settled into a soft blanket of pine needles covering a rocky ledge by the lake. There in the moonlight, Kathleen shared her experiences of recent encounters with wild wolves. We listened with great interest as her stories brought this northern forest paradise alive. Eager to give us an encounter of our own, she suggested we give a howl to see if there were any wolves nearby. She went first. From deep in her chest, she let loose one of the most realistic human imitations of a wolf I had ever heard. It sounded so good I was not even disappointed when the wolves did not howl back.

Every night after that, we all took a turn trying to speak like a wolf. Each time, we were rewarded only by the loons calling back. Although we never did hear or see wolves on that trip, we could sense that they were there. Kathleen assured us of it, but we also saw wolf scat and tracks. In addition, we learned that many of the Algonquin wolves migrate outside the park in search of white-tailed deer. Outside this refuge, wolves can become endangered by hunters and trappers. As they venture further south, even greater obstacles may threaten their lives.

On our way home from Algonquin, we drove through an area known as a potential wildlife corridor between Algonquin and Adirondack Parks. I sat quietly in the passenger seat trying to imagine myself as a wolf making the journey on foot. Exiting south from Algonquin Park on all fours, I would be home free on open land for the first leg of the trip, as there is quite a lot of forest and a few farms. But the sound of rifles might scare me: there is an open hunting season on wolves. Soon after that, I would have to cross the first major road. If I were smart, I would wait until dark, when fewer cars pass. Then I could make a run for it. That was just a warm-up. Beyond the country road, a six-lane highway packed with traffic runs from Toronto to Montreal. It would take a miracle to avoid all those cars and leap over the barriers. If I survived, I would have to find a bridge over the St. Lawrence Seaway. I could see cars driving over a suspension bridge, but it did not look as if wolves were allowed. I decided to jump in and doggie-paddle across the border into New York State. There I would only have to avoid several small towns before I arrived at my destination — the mountains of Adirondack Park. What a harrowing experience. I doubted a wolf could really make it.

Most people would prefer that wolves return from Canada on their own. After all, it is touted as "natural." The thought of capturing, drugging, collaring, and flying wolves into a new habitat seems to be a violation of the very essence of their wildness. It is much more romantic to imagine them migrating to new territories one by one, traveling up to a hundred miles in a day. I picture these wolves as pioneers, even outcasts, following a dream of new lands, opportunity, exploration, and adventure, like the first Americans who immigrated to this country.

But purely "natural" recovery may be a fantasy. Humans have already interfered with almost every natural process on earth. We were the ones responsible for exterminating wolves in the first place. We hunted them down. We destroyed and degraded their habitat. Through our roads, development, hunting seasons, and logging operations, we have made natural travel routes a deadly obstacle course. Even the St. Lawrence Seaway, which could act as a bridge for wolves when it is frozen, is also human-made. Originally constructed to facilitate commerce, the Seaway's ice is regularly broken up, making it easier for ships to travel, but harder for wolves.

The reason other types of wildlife have returned is that we have made a choice to help them. Wild turkey, deer, moose, beaver, falcon, eagle, pine marten, and others have all been given another chance by our hand. Conservation and animal welfare organizations, hunting and sporting groups, state and federal wildlife agencies, scientists and students have participated in regulating hunting, restoring habitat, protecting wildlife preserves, studying, tracking, breeding, rehabilitating, and reintroducing animals. The act of restoration is a choice — a choice to protect, to leave alone, to lend a hand. Ultimately, if we want to see a viable population of wolves recovered in the Northeast, we cannot rule out the option of reintroduction. It may just be the species' only chance. Even "natural wolf migration" is not a passive process. Ensuring the ability for wolves to safely migrate back and forth from the United States to Canada will require us to make some difficult, but necessary, choices to establish connectivity between core wolf habitats. We cannot just sit back and daydream about wolves returning home. With vision, there must be action.

Fortunately, some progress has been made. In 1997, the Wildlife Conservation Society commissioned a Geographic Information Systems (GIS) study of potential wolf habitat and connectivity in the northeastern United States and southeastern Canada. The study revealed that most of Quebec and Ontario have plenty of suitable wolf habitat—characterized by low road density and few humans. However, in parts of southeastern Canada, wolves do not seem to be expanding their territories to the south. This can be explained partially by the fact that the forests of the northeastern United States quickly give way to the farms and towns of southeastern Canada. Although the North Woods region of the United States is remote and relatively undeveloped, this same region across the border is where Canadian civilization has taken hold most intensively.

Even in undeveloped forested areas in Canada, some wolves are having difficulty. Because wolves are considered small game in Canada, they are excessively hunted and trapped. According to The Wildlands League, a chapter of the Canadian Parks and Wilderness Society in Ontario, "In Canada, some wolf subspecies may be considered *vulnerable*, such as the eastern wolf subspecies *Canis lupus lycaon*, which has lost two-thirds of its North American historic range and is frequently in contact with humans." The eastern Canadian wolf may become even more rare if recent genetic discoveries lead to a reclassification of this population as a distinct species.

As a result of lack of protection, the Algonquin wolf population, in particular, has been dramatically reduced. In the early 1990s, renowned wolf biologist John Theberge and his partner, Mary Theberge, began to track the wolves to see why the population was plummeting. They found that during winter many wolves were killed by hunters and trappers when they left the park to follow the deer. This reduced the sheer number of wolves that survived, but that was not the only ramification. Theberge's research also suggests that diminished wolf populations may affect the behavior, social cohesion, and even the tendency for wolves to hybridize with coyotes. Today, due to the Theberges' work, three counties southeast of Algonquin Park have implemented a seasonal ban on trapping and hunting, effectively expanding the protected area in which wolves can safely travel.

Increasing concern over Canada's vulnerable wolf populations and increasing interest in wolf recovery in the Northeast have prompted Canadian and U.S. conservation groups to collaborate on cross-border protection initiatives. For example, the Greater Laurentian Wildlands Project has teamed up with Ancient Forest Exploration and Research of Ontario to commission Middlebury College to map existing and potential wildlife corridors between Algonquin Provincial Park and Adirondack State Park (the area which I drove through). Known as the A to A Project, the research has revealed that a potential wildlife corridor exists along the Frontenac Axis, an undeveloped forested area between these two parks. However, trends toward increasing development, industrial forestry, and road building pose significant challenges for restoring and protecting this priority conservation zone. Removing roads and minimizing development in some areas may be necessary to restore connectivity between habitats.

The wolf populations in Quebec may have an easier migration route to Maine—there is less human development in this region—but there are still obstacles. And like the wolves of Algonquin, these animals are also experiencing similar pressures from hunting and trapping. The province's Laurentide Reserve provides habitat for the wolf population closest to Maine. Although this reserve is quite large, about 8,000 square kilometers, only thirty wolves are left within its borders. With so few animals remaining, individual biologists and conservation groups as well as a trapping association have asked the Quebec Ministry of Environment and Fauna to institute a moratorium on the trapping of wolves and to develop a wolf management plan. Unfortunately, their request has fallen on deaf ears. Not only has the Ministry refused to take action, it has also ended the scientific study of these wolves *and* continues to deny that these wolf populations are at a precariously low level.

Although Canada has thousands of wolves, the species is legally protected in less than 2 percent of its range. Therefore, the Quebec government is not legally required to take any action. Only political pressure will do the job. Several conservation groups, such as the Maine Wolf Coalition, have been working with individual conservationists in Quebec to expose this controversy. Concerned citizens from both sides of the border have begun to pressure political leaders in Quebec to protect and research this vulnerable wolf population.

For Canadians, these efforts will help protect their wolves. For Americans, increasing the number of wolves in Canada may slightly improve the chances that dispersing wolves will migrate south to find new territory. But the United States may have even more to learn from a partnership with wolf advocates in Canada. As Kathleen Fitzgerald points out, "while Northeasterners work to reintroduce wolves, the Canadians are struggling to maintain viable populations. The decline of wolves in Canada is a foreshadow to what could happen in the Northeast if we reintroduce wolves without strong protection and substantial protected habitat."

In recent years, only two wolves have been found in the northeastern United States. These animals may have migrated successfully from southeastern Canada, but there is no way to be sure where they came from. Unfortunately, both wolves were killed once they arrived in northern Maine —one in 1993, the second in 1996. Although wolves in the United States have strong protection, they can still face the same pressures as wolves that are legally hunted in Canada. Whether the threat is from poaching or an open hunting season, the only way either country will be able to truly protect these animals is to teach people to care. In Maine, the state wildlife agency has begun to educate hunters about the difference between wolves and coyotes and to warn them of the penalties for illegally killing an endangered species. To set an example, the U.S. Fish and Wildlife Service fined the first hunter and his guide who killed the wolf in Maine $4,750. Although much more needs to be done, they have made a start.

Meanwhile, scientists are also trying to determine if more wolves have migrated to the northeastern United States. Wildlife biologists conduct howling calls and filming surveys. Advocates and wildlife biologists maintain a hotline to record sightings of large wild canids. Scores of students and volunteers are eager to help by looking for tracks. Two mantras sustain this search for wolves: keep looking, keep hoping. Unfortunately, promoting natural wolf recovery is a difficult, if not impossible task. We may never know exactly what we can do to encourage wolves to migrate. It is quite possible that we cannot do a thing.

We can count on one thing, though—wolves need to roam, in search of prey, in search of mates, in search of new lands. They will roam outside national parks, outside national borders, outside the confines of our desired wolf management areas. And isolated core reserves such as Algonquin Park of Ontario or the Adirondack Park of New York cannot function as ecosystems alone. Neither the United States nor Canada can accomplish an ecologically sound wolf recovery program by themselves. We will need each other's help in a bi-national effort to protect core reserves as well as corridors between our countries so that wolves and other wide-ranging species can roam. Whether wolf recovery is accomplished through "natural" or "assisted" means, corridors between the United States and Canada will be necessary, not only for original dispersal of wolves from Canada, but to ensure the genetic, social, and evolutionary viability of the species far into the future. By establishing these connections, we also may gain a shared dream—a dream of a new land, a renewed bioregion where political and physical boundaries are torn down, where we are all united by our natural heritage.

NIGHTMARES OF WOLVES

I am walking in the woods again. Not the bright and sunny woods of a cold winter day or the lush forest of an island wilderness, but the deep, dark, scary woods of Little Red Riding Hood. *The light is dim, well past dusk. The forest smells dank from decay. The matted leaves crawl with bugs and rodents. Branches scratch my face as I try to move through the forest. Trees loom above me. I begin to run. I am lost, bewildered. I slow down to a walk.*

Where am I? I attempt to see things for what they are. I look around. I can make out steep and jagged rocks, covered with black creeping vines. I can hear branches creaking eerily in the wind. Against the dark, muddy forest floor, great white apparitions appear to be lying on the ground. I peer more closely. Then, the way you can suddenly see your reflection in the water, I see them again. This time the wolves are dead—huge white wolves, as big as the dire wolf, long since extinct. Stunned, I count. Two, five, ten, more? I bolt, and frantically search for a place to hide. As I try to crawl into the hollow under a boulder, I see one of the ghostly wolves rise to stand. The resurrected wolf is out to get me.

I am not making this up. This is the second dream I had of wolves, a nightmare really. On that morning, Wednesday, January 27, 1999, the New Hampshire State Legislature Wildlife Committee was scheduled to hold a public hearing on HB 240, a bill prohibiting the reintroduction of wolves into New Hampshire. In fact, despite significant opposition, the New Hampshire State legislature passed the law to prohibit the reintroduction of wolves into their state in May 1999. Of course, my fear was that the bill would pass. But the bill also indicated that others truly fear the resurrection of the wolf. Why?

The person behind this anti-wolf bill was a paper-mill worker in northern New Hampshire. For several months, he had been organizing protests and collecting signatures on petitions against wolf recovery. I encountered him once on September 29, 1998, at the Pinkham Notch Visitor Center in the White Mountains. There the U.S. Fish and Wildlife Service in collaboration with Defenders of Wildlife had convened a meeting to discuss the future of the eastern timber wolf in the Northeast. I saw him first in the parking lot with a sign reading, NO WOLVES! NO WOLVES! He was wearing a black-and-white striped jail suit. Later, during the meeting, he exclaimed loudly, "Wolves will keep us prisoners in our own homes!"

Clearly, his response was based on fear. A fear of the wolf, but also a fear that public support for wolf recovery was growing. His was the voice of backlash against the progress wolf advocates have made. Over the last several years, state and federal wildlife agencies have been receiving a stream of letters, phone calls, and requests from the public urging them to begin a wolf recovery program. Over 200,000 signatures have been collected on a citizen petition to support a wolf recovery study. This expression of broad-based public support as well as scientific studies indicating ample habitat and prey encouraged the USFWS to make a commitment to study the social and ecological feasibility of wolf recovery and to develop a regional recovery plan. Such a recovery plan would not mandate action. It would simply allow the public to make an informed decision about if, when, where, and how wolves could return. What is to fear?

Unfortunately, some people have interpreted "plan" to mean "scheme." They are afraid that the federal government is plotting against them. They fear conservationists are out to "lock up" the forest. They fear wolves will take away their private property rights. They fear wolves will attack their children. They fear wolves will mean lost jobs. They fear wolves will decimate deer herds. They fear wolves will prey upon their livestock.

We all know a fear of *Canis lupus*. The myth of the "big bad wolf" persists; it is rooted deep in our subconscious. It is probably the reason I had that nightmare. Yet, more and more people realize that this fear is based on fables and misinformation. Wolves certainly have much more to fear from us than we do from them. Still, wolf recovery may be seen as a loss of control over the wild. Ingrained in the New England worldview since the pilgrims landing on Plymouth Rock is the idea that progress, safety, and comfort depend upon domination and domestication of all things wild. To those who still believe this, the resurrection of the wolf may symbolize a haunting reversal of progress.

The truth is that there is not one good reason to fear the wolf. Evidence from experiences with wolf recovery efforts in other parts of the country show that wolves do not pose a significant threat to people's livelihoods, livestock, children, traditional hunting rights, forest practices, or deer populations. But given the prevalence and durability of these fears, ultimately a successful wolf recovery program will take nothing short of a paradigm shift. The facts are on the wolf's side. However, sometimes, dry facts are not enough. Open minds, critical thinking, education, and time are what it will take.

Protecting habitat for wolves and other species is not merely an ecological challenge; it is a social one. The most straightforward approach to wolf recovery may be to draw a line around a protected core habitat or even provide incentives for private land owners to allow wolves on their land. But the protection of wolves in the broader landscape will require a cultural acceptance and appreciation of the animal by the people who live, work, and recreate in the North Woods. Negative human attitudes toward the wolf are one of the greatest challenges to successful wolf recovery.

Although opinion polls indicate that the majority of the public in the Northeast favors the wolf's return, many individuals simply loathe the animal. These are the people who make wolf advocates nervous. Lamentably, these are also the people who get all of the media attention. It only takes a few individuals to wipe out a small population of wolves. In New Mexico and Arizona, for example, eleven Mexican wolves were reintroduced in the spring of 1998. By November, five of these wolves were found shot to death, probably as an act of revenge by one or two wolf-haters.

Such cruelty is profoundly discouraging. However, can we wait for every single person to accept the wolf before a wolf recovery program should proceed? Even if wolf advocates had all of the time and resources in the world, it would be impossible to encourage everyone to change their mind, especially the most avid anti-wolf zealots. In the Southwest, the wolf killings did not cause the USFWS to abandon the project. Instead, the program persevered in the face of adversity. Wildlife biologists replaced the dead wolves with fourteen others, and plan to reintroduce one hundred more animals over the coming years. As wolves become a more permanent and natural part of this landscape, it is hoped that the animosity toward them will cool down.

Perhaps it will take the actual presence of wolves in their former habitats, the forests that people know and love, to really shift people's attitudes. It is easier to accept what is there, the known, rather than the unknown. As long as the wolf only exists in our fairy tales and dreams, people can imagine the creature to be as frightening as they choose. If the majority of the people in the Northeast want wolves back, and so far it looks as if the numbers are on the wolf's side, a few individuals cannot be allowed to ruin the process for everyone else, ruin it for the wolves. This is, after all, a democratic society.

Yet, we cannot ignore the threat that humans pose to wolves. If we choose to reintroduce the species into an area of the country with a dense human population nearby, we have a responsibility to teach people how to coexist with the wolf as well as provide safe havens where they are secure from human threats. Successful coexistence will take more than a begrudging tolerance, it will take people who welcome, even celebrate, the wolf's return.

Once people learn to understand the wolf, they often learn to love the animal as well. That is why education is one of the strongest and most effective tools of wolf advocacy. In the past five years, wolf education programs have proliferated in the Northeast. Hundreds of teachers in elementary and secondary schools have chosen the wolf as a subject to study in science, English, and history. Educational groups such as the New Hampshire Wolf Alliance and Vermont Institute of Natural Science teach children and adults about wolves so they may have a greater understanding and acceptance of all wildlife. The Maine Wolf Coalition has opened a year-round educational wolf center in Hallowell, Maine. Exchange programs bring hunters, loggers, and farmers from other wolf recovery areas to share their experiences of what it is like to live near or in wolf habitat. Several major conferences have been held in the Northeast to share the latest and best thinking on the implications of wolf recovery. In communities across the region, wolf advocacy groups present slideshows to thousands of people at rotary, sporting, and garden clubs, religious associations, at country fairs—wherever people are willing to listen and learn.

An unfortunate consequence of our culture's growing affection for wolves is that many people mistakenly believe these wild creatures can be pets, just like the family dog. Usually this relationship results in disastrous consequences. It is very, very difficult to make a wolf happy in a house or backyard. As a result, when wolves grow up from cute little puppies to 100-pound adults, they are abused, abandoned, or given to shelters where, at least, they can live out their lives. In an effort to find a positive side of this bad situation some of these wolves have been chosen as "ambassadors" to the world of humans and provide an exhilarating focus for education.

Several ambassador programs tour or are based in the Northeast, including Wild Sentry, from Montana; Mission: Wolf, from Colorado; New York Wolf Center; Wolf and Wild Canine Sanctuary, in Vermont; Loki

Clan Wolf Refuge, in New Hampshire; and Wolf Hollow, in Massachusetts. Each of these programs is different, but they have one thing in common: the living presence of a wolf. Wild Sentry, for example, is led by a husband-and-wife team, Pat Tucker, a wildlife biologist, and Bruce Weide, a storyteller. Through their slide presentation they provide clever facts about wolf biology and tell enchanting myths and fables. But the dog and wolf team, Indy and Koani steal the show. Indy, a lovable mutt, wanders in and out among people's legs during the presentation, like an excited little kid eager for attention. Koani, on the other hand, is described by Pat and Bruce as a mature adult who is often quite bored by the whole ordeal. For the first time, many people can see for themselves who the wolf really is. Before their very eyes the animal transforms from a fang-toothed monster to a noble, if aloof, wild creature.

In just a few decades, Americans have developed a fascination with the wolf. Certainly, there are still people who despise the animal. But even most opponents of wolf recovery will admit that they have nothing against the wolf itself. Countless movies, documentaries, books, tee-shirts, artwork, music, even comic strips have glorified this wild canine. If, in some distant time, a cultural anthropologist finds these twentieth-century artifacts, they may surmise that Americans worshipped the wolf as a god. But, if they were to look closer, a little further back into our culture's history, they would find something else—artifacts showing a war on the wolf. A keener observer may see that our culture is really making amends. After centuries of ignorance and hatred toward the animal, we have changed our minds. We celebrate the wolf because we need to, because we need to redeem our own species as one that respects and helps sustain all forms of life.

A DREAM OF DEMOCRACY

I am standing on a stage in front of a boisterous crowd. Three hundred people or more file into the school gym. Some knock over the bright red plastic chairs as they make their way to their seats. Others sit patiently with their notebooks. All sorts of people are here—upstanding townsfolk, rednecks, hippies, yuppies, farmers, land owners, conservationists, wolf lovers, nature lovers, animal lovers, wilderness activists, politicians, business leaders, students, parents, loggers, hunters, hikers, scientists, bureaucrats. The meeting commences, and people start to voice their opinions about the wolf recovery program. For a second, the noise is deafening. Then I begin to distinguish what people are saying. I am startled by who is saying what.

A hunter stands up and makes a passionate plea for wolves, calling them "the great hunters." A mother with an infant in her arms presents two thousand petitions for the wolf, names she has collected in her spare time. A logger says wolf recovery won't threaten him because wolves don't eat trees. A land owner wants everyone to know that he would welcome wolves on his land. A sixth grader painstakingly explains how wolves will improve ecosystem health and enhance biodiversity. The mayor holds up a dollar bill to show that wolf recovery will bring in money to local economies. A politician cries out, "wolves were here first!" The wolf advocates just sit there silently, smiling at all the others who are speaking for the wolf.

Sound too good to be true? Well, it was only another dream. But I don't think it is that far from the truth. Just eight years ago, the idea of wolf recovery was deemed impossible. Now, we are already succeeding beyond our wildest expectations.

When people ask me what I do for a living, I tell them I am a "wolf advocate." The next question, "A what?" is often accompanied by quizzical frowns or confused chuckles. Who are wolf advocates anyway? Why do we do what we do? How many of us are there?

Individuals and organizations choose to advocate for wolf recovery for many different reasons. Often these concerns reflect individual personalities and perspectives. Other times, they are built into the professional goals of an organization. There are perhaps as many motivations to become a wolf advocate as there are people who care. Each rationale is as good as another. But sometimes these justifications themselves are vulnerable to criticism from those who oppose wolf recovery. Only when united can these reasons build the strongest case for the wolf.

For many wolf advocates, the driving force is a personal connection to the wolf—an emotional affection for the animal. In conservation circles, wolves are referred to as "charismatic megafauna," powerfully attractive large mammals that are known to elicit powerful emotions. But there may be more to it than that. In some Native American cultures, wolves are a totem—an animal that is considered kin, a teacher to guide one through life. The wolf's life force is hard to disregard. Indeed, most human cultures in the Northern Hemisphere have been attracted to the wolf. The proof is in our long-standing relationship to the canine—in the form of "man's best friend," the dog. We have gone to great lengths to domesticate wolves for special traits—companionship, strength, loyalty, passivity, a nose for hunting. Depending on the climate and cultural demands, a variety of dogs have developed by careful selection and breeding. The Chinese chow, Mexican Chihuahua, German dachshund, and the good old yellow lab are each as genetically similar to the wolf as the next. In fact, the Latin name for the domestic dog is *Canis lupus*, the same as the gray wolf. While most dogs have become incredibly ill adapted to the wild ecosystems from which their ancestors were taken, our friendly neighborhood canines are a living testimony to how important the relationship between human and wolf is.

Wolf restoration is also motivated by moral convictions. For many people, the oppression of wolves is thoroughly unacceptable and unjust. What Americans did to the wolf was fundamentally wrong. The war on the wolf, an extermination campaign that lasted for centuries, can be compared to acts of genocide against Native Americans, Jews, and other persecuted minority groups. For those who empathize or identify with this level of persecution, wolf recovery can take on ethical and religious dimensions. To push a species over the edge of extinction may be the ultimate sin. To return the wolf to its rightful home may be the only way we can finally make amends.

For the conservation-minded, the wolf is defined as a vital part of the North Woods ecosystem. Life on earth can no more do without earthworms, insects, or trees than without the wolf. The wolf itself is considered a keystone species—an animal that helps maintain the biodiversity of the entire web of life. To restore the wolf is an ecological imperative. It is a mandate informed and guided by science. Indeed, the wolf is protected by the Endangered Species Act for one reason alone—the American people have a responsibility to save the species from extinction on our own nation's ground.

Finally, wolves can be a powerful catalyst for wilderness restoration. The howl of the wolf echoing across the forest is an unparalleled inspiration. The act of a wolf pursuing its prey is an ancient ritual, the essence of wild natural processes. The close familial bonds within a wolf pack emphasize the intimate interconnections that all wildlife depend upon. Wolves bring us hope. If we can manage to restore healthy, viable populations of free wild wolves, then we might have the wisdom it takes to restore wilderness as well.

All of these are compelling motivations for individuals to get involved in northeastern wolf recovery. Indeed, thousands of citizens have already been inspired to take a stand. During the seven years since a wolf recovery movement has taken hold in the Northeast, grassroots support for the effort has grown like wildfire. From a few vocal individuals, to one organization, then two, then four, ten, and more. Today, over thirty conservation organizations have become actively involved. In all, these groups represent hundreds of thousands of constituents.

Recently, these conservation groups formed the Coalition to Restore the Eastern Wolf (CREW) so that we could work together toward a common goal. That goal is the recovery of viable populations of the wolf in as much of its former range in the northeastern United States and southeastern Canada as is feasible. Each of the organizations in CREW brings a unique contribution. Some organizations bring strength in numbers, joining thousands of members through networks at meetings, conferences, and the Internet. Other groups influence the movement with their vision, urging people to look beyond perceived political constraints to focus on restoring wild and free wolves to a wild land. Others unite scientists with activists, firmly grounding advocacy in the biological and ecological needs of the species. Experience is essential too. National groups bring their expertise and contacts from other wolf recovery areas around the country to help this region navigate through a complex process. Then there is the authority that comes with being rooted in a specific place. State and local groups can build solid relationships with their wildlife agencies and inspire the citizens of their state to lead the way toward wolf recovery. Together, these organizations offer a wealth of experience, knowledge, dedication, and resources. Still, we cannot succeed by ourselves. CREW invites people and organizations of all types to get involved.

The future of the wolf depends upon many more citizens, conservationists, land owners, hunters, scientists, teachers, farmers, parents, children, and a host of other constituents to take up the call. It will take years of planning by federal and state wildlife agencies as they develop a blueprint for wolf recovery. It will take dozens of deliberations with stakeholders and scientists. It will take hundreds of public hearings, thousands of letters, and pages and pages of studies. It will take pressure and perseverance. It will take wisdom and understanding. It will take courage. It will take time.

Ultimately, the wolf's fate will be decided in the meeting houses and halls of democracy. And democracy can be messy. Animated by passions, our citizens' voices are often loud and hard to listen to. But this is where we must place our trust for wolf recovery to succeed. Naysayers caution that wolf recovery is too charged with controversy and animosity, and certainly we have a long way to go. People feel strongly about wolves. They either love them or hate them. For many people, the wolf is a symbol for something else—for federal government, for the greens. But stereotypes and symbols can box us in. They can limit our dreams of what is possible, our expectations of who will say and do what. We need a variety of people involved, not just because it is the right thing to do, but because many more people out there, beyond just "wolf advocates," want to make this dream a reality.

Wolf recovery is not about us versus them. It is about creating a culture of acceptance and generosity, not only for the wolf and other wildlife, but for each other. It is about recognizing the rights of the wolf, as one of the original citizens of the North Woods. It is about listening, communicating, informing, and empowering one another, so we can make wise choices. It is about cultivating the conditions in our heart and in the land to sustain all things wild. It is about dreaming a big enough dream to restore and heal a broken landscape.

WOLVES DREAM TOO

I am sure wolves dream, too. You have seen your dog lying asleep on the floor, kicking and whimpering. What do you think dogs dream of? I think wolves dream of home—the deep green forests where moose, beaver, and deer live. They dream of running, miles and miles each day. They dream of playing with their pups. They dream of deep snow, strong winds, and clear sunshine. They dream of loosing a long howl. They probably don't dream of us as we dream of them. But, perhaps they know, somehow, that humans in the Northeast are calling for them to return home; that together, we can call back the soul of the wild.

Appendix A
Coalition to Restore the Eastern Wolf (CREW)

The Coalition to Restore the Eastern Wolf (CREW) comprises thirty-seven organizations that support the recovery of viable populations of the wolf in as much of its former range in the northeastern United States and southeastern Canada as is feasible. Toward this end, we are working to:

1. Encourage the U.S. Fish and Wildlife Service to complete and implement a wolf recovery plan that provides for a timely analysis of all factors relating to the potential for recovery of the eastern timber wolf in the Northern Forest of Maine, New Hampshire, Vermont, and New York. Along with the recovery plan, an environmental impact statement (EIS) or environmental assessment (EA) should be done in partnership with the four states and ensure full public review and participation;
2. Ensure the continued protection of wolves in the Northeast under the Endangered Species Act (ESA) for as long as necessary to ensure their long-term recovery;
3. Promote cooperative efforts between the United States and Canada to protect existing wolf populations and facilitate travel by wolves between the two nations; and
4. Undertake research and educational efforts to facilitate public awareness about wolves and wolf recovery.

Participating organizations include:

Adirondack Council (NY)
Biodiversity Legal Foundation (CO)
C.L.A.N.: Wolf Defenders (Quebec)
Defenders of Wildlife (CD, NY, ME)
Environmental Advocates (NY)
Forest Ecology Network (ME)
Forest Watch (VT)
Greater Laurentian Wildlands Project (VT)
A Hunter's Voice (OR)
Institute for Environmental Learning (NY)
Keeping Track (VT)
Loki Clan Wolf Refuge (NH)
Maine Wolf Coalition (ME)
Mission: Wolf (CO)
Mt. Kearsage Indian Museum (NH)
Native Forest Network (VT)

National Wildlife Federation (VT)
New Hampshire Wolf Alliance (NH)
New York Wolf Center (NY)
Northeast Ecological Recovery Society
  (NY)
North American Wolf Foundation (MA)
Northern Appalachian Restoration Project
  (NH)
Predator Project (MT)
RESTORE: The North Woods (ME & MA)
Sierra Club—New Hampshire Chapter
Sierra Club—Maine Chapter
Sierra Club—Atlantic Chapter
Sinapu (CO)
Southern Appalachian Biodiversity Project
  (NC)

Vermont Institute of Natural Science (VT)   Wild Earth (VT)
The Wildlands League (Ontario)   Wild Nature (NH)
Wild Canines Unlimited (NH)   Wild Sentry (MT)
Wild Dog Foundation (NY)   Wolf and Wild Canine Sanctuary (VT)

For more information contact: Kristin DeBoer at RESTORE, 978–287–0320.

Appendix B
Northeast Wolf Recovery: Timeline of Major Events

1992

- U.S. Fish and Wildlife Services publishes a revised version of the *Eastern Timber Wolf Recovery Plan* and identifies the Northeast as a potential wolf recovery area.
- Citizen activists begin a public dialogue on wolf recovery by publishing letters to the editor in various newspapers around the region.
- RESTORE: The North Woods launches the first campaign for wolf recovery in the Northeast.

1993

- A bear hunter shoots and kills a confirmed gray wolf in northern Maine. He and his guide are fined $4,750 for illegally killing an endangered species.
- A wolf sighting hotline is established.

1994

- Several new grassroots wolf groups form, including Maine Wolf Coalition, New Hampshire Wolf Alliance, Wild Dog Foundation, and Northeast Ecological Recovery Society.
- Wolf educational organizations, such as Institute for Environmental Learning, Mission: Wolf, Wild Sentry, and Wolf Hollow, begin to incorporate message of Northeast wolf recovery into their programs.

1995

- Defenders of Wildlife convenes two meetings on wolf recovery at the Roger Williams Zoo in Rhode Island and Wolf Hollow in Massachusetts.
- The Greater Laurentian Wildlands Project begins to work with Canadian groups to identify a potential wildlife corridor for wolves between Algonquin and Adirondack Parks.
- RESTORE: The North Woods proposes a 3.2-million-acre Maine Woods

National Park that would provide a sanctuary for wolves and other large predators.

1996

- A trapper kills a "probable" gray wolf in downeast Maine.
- Wolves of America Conference sponsored by Defenders of Wildlife is held in Albany, NY.
- Maine Department of Inland Fisheries and Wildlife begins to conduct wolf tracking surveys and to educate hunters on the difference between coyotes and wolves.

1997

- The Coalition to Restore the Eastern Wolf (CREW) is officially formed, comprising thirty-seven local, regional, and national organizations that support the recovery of the eastern wolf.
- Wolves in our Backyard Conference sponsored by Wild Canines Unlimited and Antioch New England Graduate School is held in Keene, NH.
- New Hampshire governor Jeanne Shaheen signs a proclamation declaring a Wolf Awareness Week.
- Wildlife Conservation Society releases a study done by Daniel Harrison and Theodore Chapin on a *Potential Dispersal Corridor for Wolves Between Laurentides Region, Quebec and New England*.

1998

- U.S. Fish and Wildlife Service announces its intention to develop a regional wolf recovery plan for northern New England and New York at meeting in Pinkham Notch, NH. Protestors vocally oppose wolf recovery.
- Wolves and Human Communities Conference sponsored by The Hastings Center is held in New York City.
- Maine Governor Angus King proclaims and then, realizing his mistake, rescinds Wolf Awareness Week.
- Alarmed that the wolf population in the Laurentide Reserve of Quebec has dwindled to only thirty animals, biologists, trappers, and conservation groups call for a three-year moratorium on wolf trapping.

1999

- CREW convenes four state-by-state meetings with the wider conservation community to discuss the future of Northeast wolf recovery.
- The State of New Hampshire passes a law prohibiting wolf reintroduction in that state.
- Maine Wolf Coalition opens its new Northeast Wolf Center in Hallowell, Maine.
- The Sportsmen's Alliance of Maine tries to introduce an emergency bill to prohibit wolf reintroduction into Maine, but fails because of support for wolves.

- Conservation Biology Institute releases a Defender's-sponsored study on the feasibility of wolf reintroduction in Adirondack State Park.
- Across the Northern Forest, multinational paper corporations begin to sell off large tracts of land within potential wolf habitat. A small portion of the land is protected from development by private land trusts and state governments.

2000

- The Missing Link?: Prospects of Wolf Recovery in the Northeast Conference sponsored by National Wildlife Federation is held in Brunswick, Maine.
- The Vermont State Legislature tables a bill to prohibit the reintroduction of wolves into that state.

LITERATURE CITED

Berry, Thomas. 1988. *The Dream of the Earth*. San Francisco: Sierra Club Books.

Chadwick, Douglas. 1990. *The Kingdom: Wildlife in North America*. San Francisco: Sierra Club Books.

Gaillard, David et al. 1999. "At a Crossroads: The Wolf and Its Place in the Northern Rockies." Montana, The Predator Project.

Harrison, Daniel J., and Theodore B. Chapin. 1997. "An Assessment of Potential Habitat for Eastern Timber Wolves in the Northeastern United States and Connectivity with Occupied Habitat in Southeastern Canada." Wildlife Conservation Society, Working Paper no. 7.

Leopold, Aldo. 1949. *A Sand County Almanac*. New York: Oxford University Press, Inc.

Mills, Stephanie. 1996. *In Service of the Wild: Restoring and Reinhabiting Damaged Land*. Beacon Press.

Mladenoff, David, and Theodore A. Sickley. 1998. "Assessing Potential Gray Wolf Restoration in the Northeastern United States: A Spatial Prediction of Favorable Habitat and Potential Population Levels." *Journal of Wildlife Management* 62(1):1–10.

National Wildlife Federation and The Rainforest Alliance's Smart Wood Program. "DRAFT Smart Wood Certification Program: Northeast Guidelines for the Assessment of Forest Management." Montpelier, Vermont.

Paquet, Paul, et al. 1999. "Wolf Reintroduction Feasibility in the Adirondack Park: Prepared for the Adirondack Citizens Advisory Committee." Corvallis, Oregon: Conservation Biology Institute.

Peterson, Rolf O. 1997. "Ecological Studies of Wolves on Isle Royale: Annual Report 1996–97. Houghton, Michigan: School of Forestry and Wood Products, Michigan Technological University.

RESTORE: The North Woods. 1999. "America's Next Great National Park: Preserving Our Maine Woods Legacy."

Theberge, John B., and Mary Theberge. 1998. *Wolf Country: Eleven Years Tracking Algonquin Wolves.* Toronto: McClelland and Stewart.

Thoreau, Henry David. 1864. *The Maine Woods.*

Trombulak, Stephen, and Jeff Lane. 1996. "A Proposal for a Priority Conservation Zone in Northern New York." Middlebury, Vermont: Middlebury College.

U.S. Fish and Wildlife Service. 1992. *Recovery Plan for the Eastern Timber Wolf.* Twin Cities, Minnesota.

U.S. Fish and Wildlife Service. "Red Wolf Restoration Program." Unpublished paper. Manteo, NC.

Wessels, Tom. 1997. *Reading the Forested Landscape: A Natural History of New England.* Vermont: The Countryman Press.

White, William M. 1982. *Sweet Wild World: Selections from The Journals of Henry David Thoreau Arranged as Poetry.* Charlestown, Massachusetts: Charles River Books.

The Wildlands League. 1997. "Kill the Myth, Not the Wolf: A Wolf Conservation Strategy for Algonquin Park and Region." Toronto, Ontario: An Algonquin Park Watch Publication.

Vermont as Montana

Rick Bass

WE KILLED ALL THE WOLVES. WE GOT ALL THE caribou, all the wolverines, all the lynx, and all the grizzlies. In the East, all the elk. Anything that was grand or magnificent or even rare, we got it. But we weren't finished. We turned our murderous, frightened instincts elsewhere—upon voiceless ones. Anything that could be controlled or dominated, anything meek, we went after with savagery. A person can argue all day long about whether humans are filled with light or darkness, or varying percentages of both, but what I think is inarguable is that we are insatiable. We are almost always creatures of hunger.

And now we want wolves back in the world again. Or rather, some of us want wolves. A part of our culture, much of it highly urban, hungers intuitively for a wildness, a biological integrity, that has no name, but that has been sculpted into abstract icons by both the Madison Avenue marketeers and the mainstream environmental movement. These icons possess the visage of the wolf.

I think two dynamics are at work here. Part of our longing for wolfishness is spiritual—hungering for the kinds of grace and wildness that are increasingly being diminished by our own clumsy presence. But part of it, I fear—the hunger for wolfishness—is being manipulated.

The basic question in the media, government, and communities both urban and rural, is: *Do we want wolves?* Any poll that you care to look at responds with a resounding yes.

To me, however, more interesting questions exist, which are not currently being asked: *Why do we want wolves?* and, perhaps unanswerable, but worthy of consideration: *What would the wolves want?*

How badly do they desire to be in the places where we want them to be? How badly does the land desire them, in those places, and how capable is the land of supporting the force of wolves—and the surrounding human communities?

We want them in Montana, Wyoming, Utah, Idaho, and Colorado. We want them in Oregon and Washington, and we want them in Arizona, Texas, and New Mexico. We want them in Louisiana and Mississippi, and in the Carolinas and Kentucky. We want them back in Mexico, and in Alaska, British Columbia, and the Yukon, and in Minnesota, Michigan, and Wisconsin.

And we want them in the Northeast, where (as with all those other places, and more), they once lived, before we told them to go away: before we trapped, poisoned, and killed every last one of them.

We want them in Maine's North Woods, and in the vast landscape of the Adirondacks (even as acid rain is scouring the highest peaks and stony ridges, the forests and the mountain lakes). We want them up in Vermont's Northeast Kingdom, and there are even those among us who want them in the Berkshires of Massachusetts.

Is not such hunger unseemly—almost avaricious?

Like so many others, I want them back in these places, too. But I often get the sense that I want them there under slightly different conditions than many of my other environmentalist friends, or even the "pro-nature" constituency that comprises the majority of this country.

Perhaps I'm a spoilsport, but I fear sometimes that we are going about this in the wrong manner, as we pursue our desire. I fear that we have become so acquisitive in our shop-'til-you-drop culture and in the milieu of instant gratification that this is the pattern that is developing with regard to the recovery of endangered species: we borrow some from one place and transport them to the new place, as one would go into a store and browse the sale racks. We achieve within a year or two our desire, our goal, but then all but abandon our attention to the thing that we were only beginning to try to learn to understand.

I worry about the fit.

I worry about the loss of custom-fitted, evolutionary, earned grace, wherein a species carves its way, with integrity and sinuosity, into a healthy, dynamic ecosystem—and then endures, with those same qualities.

I know I sound like somebody's old grandfather, carping about the process, rather than being all-enthusiastic about the thing itself. But an inescapable question for me has always been: If you airlift a wolf out of Quebec and set it down in a different habitat, is it still a wolf? Well, yes, of course, but—

Once we have achieved the airlift and satisfied our desires, once the wolves are running again through the Northeast, will we, bored, turn away? Or will we commit to protecting and preserving the other, less glamorous issues the wolf needs for long-term survival, chief among them an acknowledgment by us that we do not have all the answers? Is there still a chance that we might yet shed the old and broken belief that nature, especially the wilder reaches where we have not yet imposed too heavily, is always a chaotic, awkward accident in need of repair, while we are the designers, the keepers of the garden?

Maybe my anger or frustration is simply that it has come to this. When I consider the resident wolves of another nation—entire families of wolves—being harassed and pursued, then captured and flown by helicopter to a country not their own, some inarticulate clamoring, some violent protest arises in my blood against my own views of wildness.

Am I being a little too pure-minded about this? Why do I respond so strongly to this short-circuiting of the evolutionary process?

In the kind of perfect world a writer can try to create on paper, we the people would look past the glamorous, athletic grace of the wolf, past the anthropomorphic characteristics of the pack, past the cuddliness, past the loyalty and power, and instead try to see the landscape as a whole—a complete landscape as full of yearning and design but also chaos as any other living, breathing thing.

Instead, I get the sinking feeling that—aided and abetted by the marketeers of glamour-capitalism, and by the sheer weight of our culture—we have too much tendency to look at the wolf, or any other singular species, as a component: a replaceable part. Was it only a hundred and fifty years ago that the first interchangeable parts were designed for the cotton gin; less than a hundred years ago that a similar concept was applied in the manufacture of automobiles—a move that relied dramatically on the division of labor into many parts that, though each was specialized, were easily replicated?

In retrospect, it shouldn't seem so shocking. And I'm not saying it's evil, or even wrong—replacing a craftsman with a laborer, and then, later in the century, replacing a laborer with a machine.

Anyone should have been able to see it coming. If we could treat humans as replaceable parts, then how else could we be expected to view wolves, grizzlies, condors, or bison—much less moths, salamanders, beetles, fungi—in any terms other than a checklist that simply denotes *absent* or *present*?

In a perfect world—the one I want—we would still be clamoring for the return of wolves to the ecosystems from which they were extirpated. But we would also be demanding, with equal resolve, for the protection and preservation of mystery, and of the last few wild landscapes we still have remaining.

These remaining islands of wilderness are now but the faintest residue of who we once were as a culture. They are not so much the part of the meal that we haven't yet been able to eat, but rather, the last morsels tucked away in crevices so remote that we have simply not yet been able to get to them. These lands can still be protected, and I am not yet ready to cede their loss to all the other vanished or consumed landscapes. I am impatient for us to begin reassembling these old islands, and reconnecting them into some braided weave of strength.

Sometimes I find myself believing that time is neither linear nor even circular, but elliptical; that it moves across the landscape and through our brief lives like the movements of an animal that travels through the forest for a while, then saunters, then perhaps lies down and sleeps, before rising again—wandering, hurrying, resting—and that a chart of these movements would reflect a pattern, a rhythm, more like the tracks of a snake undulating across sand, than the analog measurements of steady, incremental circumstance.

In any event, we can't go back in time, nor do I really think we can change who we are, *the hungry ones.*

We can try to address the exploding human population; but we know realistically that both the mass and the momentum of this force are asteroidal in scope, and a problem that generations will almost surely be addressing for a long time. It's far beyond our immediate control.

What is not yet beyond our control is to see our last few wild ribbons of landscape for what they are—and to work to protect them, and to honor them as the sculpted pieces of creation that they are.

With no disrespect to the wolves, I am impatient for us to look past the wolf, over the wolf's shoulder, at the landscape beyond. The issue of wolf reintroduction is largely political now, and one day soon wildlife management may be merely a matter of politics—but for now ecology and landscape are at least a consideration.

Once upon a time they were everything.

With all due respect, as marvelous as wolves are, they're relatively easy to recover, if our society continues to change its mind and decide that we need or want them after all. Wolves are what biologists call "plastic"; they're able with skill and hunger and intelligence to squeeze into available niches often simply by virtue of their great force. As long as there is something for wolves to chase down and eat (and it can be almost anything), then wolves will have a chance. The chief factor in their survival is the need to have people stop shooting at them.

I believe it's the preservation of the wild forests, the wild ecosystems— and, in many instances, their restoration and recovery—that will be our greater challenge.

I live in a wild valley in northern Montana. Wolves are infrequently sighted in my valley—not every year, but some years, by a few people. In thirteen years I have seen wolves four times, have heard them howling eight times, have dreamed of them twice, and have seen their tracks fewer than a dozen times.

It is always a profound pleasure to come across their tracks, and I have come to like the fact that the valley's wolves are both so invisible and so nomadic. One of the best hunts I had last year involved my tracking a herd of nervous elk way up high. I never saw the elk—I came upon their tracks and then discovered where they'd spooked and run, which puzzled me at first, since I'd come in from downwind. But a short while later I found the tracks of one wolf, upwind, who'd been following them, walking along behind them, just following. He was the reason they were so skittish. His tracks indicated he never gave chase; he'd just been following. I never saw him, or the elk.

It was cold, windy, high country up there, with a beautiful view of the valley, and it pleased me to know that there had been both elk and a wolf on that mountain only a few hours earlier. That such simple but wonderful happenings were not yet confined merely to history books and tales of long ago.

For a long time I used to worry that one of those infrequent wayfaring wolves would tangle with one of the two dozen or so cattle that are kept on small pastures in the river bottom clearings in this northern forest: two dozen cattle amid almost a million acres. (Another hundred head or so are run up in the northeast corner of the valley, in the mountainous backcountry, scraping out their brief existence prior to the slaughterhouse by gnawing at the weeds and clover that grow alongside the myriad logging roads that tunnel their way through the densely forested mountains; but such is the lot of those wild, turned-out wastlings, in a land of grizzlies, mountain lions, coyotes, wolverines, runaway logging trucks, poachers, and all else, that wolves would be the last in a long line of suspects—and rightly so— were any of those cattle roaming the national forest to disappear.)

It's those two dozen cattle down on private land in the valley's interior that I used to worry about, knowing that if, even across twenty-five years' time, one wolf ate one cow—four cows per century—it could spell doom for any subsequent wolves attempting to recolonize this valley. That people might decide, *Aha, wolves are bad.*

But one day I finally stopped worrying. I'm not saying it couldn't happen yet. But eventually I began to look at the situation from a wolf's perspective.

In addition to those two dozen cattle huddled down by the barns of men, surrounded by the open spaces and barking dogs and road traffic and tractor-grumble and chimney-smoke and firearms, there are perhaps a thousand elk, a thousand moose, and ten thousand deer, all of which taste better and are far more safely obtained.

As ever, I had put our species and our concerns at the center of reason in the universe.

We are an immense, ever-increasing force in the universe (not as immense as bacteria, or insects, or even the daily pulses, the respirations and transpirations, of the breathing forest—but nonetheless, huge). Our hugeness, however, in no way means that all the rest of the world adjusts and orders itself accordingly to our desires. The tides do not notice us, nor do the seasons, nor the rain, nor geology.

This is a lesson most of our species learns somewhere between kindergarten and the end of high school. But then a funny thing happens. We go out into the workplace, or into the marketplace, and it is as if we enter a second childhood, wherein the world once more shifts and reorganizes it-self, with our goals and desires as well as our fears once more acting as the central pivot point.

Again and again, it usually takes a good long walk in the woods to re-mind us of how the world really is and that, as far as grace is concerned, we are really only fringe players.

I can't begin to count or consider all the reasons I believe it is impor-tant to keep wolves in our forests, or to help them recover in the places where they once were, and are capable of re-inhabiting. But neither can I escape the moral queasiness that, with the hungry zeal that is a hallmark of our species, we might overdo it and try to mash wild wolves into places and situations where we are not yet ready for them. Places where we are not yet committed to understanding, or, if not understanding, at least revering.

At any rate, one thing I'm not that worried about any more are the two dozen cattle in my million-acre valley. Thanks to the vision of a group called Defenders of Wildlife, in the unlikely event that these nearly invisible wolves did take a cow, the rancher would be paid for the value of that loss, as long as it could be proven that the wolf was indeed the predator. (This isn't always possible, but for now, it's the best the system can do. It's cer-tainly a far cry from the old days.) And I'm not saying it won't, or doesn't, happen. It does, and will.

In Montana, where we currently have about a hundred known wolves, the winter of 1997/98 was the worst on record, and led to a huge die-off in the deer herds. This in turn sent predators down lower out of the moun-tains searching harder for the surviving prey, where the predators were more likely to encounter domestic stock.

In that one exceptional year, wolves killed almost sixty livestock animals, which sounds like a lot, and is a lot, if any of them are your animals. But it should be mentioned also that each year in Montana about 170,000 stock animals die before being taken to market for us to buy and eat them, and that on a ten-year average, wolves kill only about 6.5 livestock per year. Michael Jamison, a reporter for the *Missoulian* newspaper, writes, "That's to say they [wolves] killed one in every 25,000 animals that died before go-ing to the butcher." Domestic dogs, Jamison reports, killed 1,600 animals in 1995. Mountain lions took another 800, and coyotes "killed a whopping 29,000 head of livestock in 1995 . . . That same year, wolves killed 4."

It's well-known that wolves displace other wild canids such as feral dogs and coyotes from their territory. So wouldn't it be *good*, in the long run, to have wolves in an area?

Somebody is not stopping to look at the ecology of the situation.

Already—even when armed with the science, the data—we are not looking past the wolf. We're not looking ahead, and we're not thinking or planning. We're still just reacting.

As with the grizzly, and nearly every other species on our shamefully long list of endangered species, the larger question of habitat destruction, particularly wilderness destruction, has been ignored. Instead, the animals —ultimately unknowable, glorious, world-crafted, and vastly superior in grace—are quickly fashioned by and into omens and symbols. Many ranchers, perhaps most, don't hate the wolf. Instead, the vitriol usually arises from the right-wing advocates of increased property rights, and the bought-and-paid-for advocates of increased corporate liquidation of the public lands in the West, who in addition to failing to see the forest, fail also to even see the wolf, viewing it instead as only an emblem of all else that they hate, which is in itself a long list.

In their minds, it all gets mixed together, so that the green-eyed wolf represents not just the government (the IRS, FBI, DEA, CIA, OMB, etc.) but also the fear and anger that their land will be confiscated, once wolves return: a fear that has no historical precedent, but which lives large in them nonetheless.

Let me hasten to point out that environmentalists are no pure lilies at this game, either. It's often unavoidable: with a creature as complex and magnificent, as visually appealing as the wolf, we can't help but load onto the wolf's back not only our fears but also many of our goals and desires— the hungers and needs—that we want associated with such a powerful, attractive representative.

How strange it might seem, at the outset, to begin a discussion of the return of wolves to the Northeast by talking instead about this remote cutover valley in northern Montana, where perhaps 95 percent of the valley's human residents have never seen a wolf; a valley, a landscape, through which one or two wolves might drift every one or two years. But I think inextricable from this discussion is one central question, which is, *How much of what is wild are we going to attempt to protect, and take with us into the future?*

I may commit a skip of logic here, please allow me. I do not mean to fall into the trap of equating the wolf with wilderness, because not all wolves, all of the time, need or perhaps even desire wilderness. They need only to be able to live in places where they will not be persecuted by humans, and where deer or other game abound.

Often this kind of habitat is found in a tenuous middle seam between our high-altitude rock-and-ice regions of "designated" wilderness—where, due to the harsh and rocky terrain, deer are sometimes in short supply—and the rural villages and hamlets, along valley river bottoms, where deer are plentiful.

In a perfect world, the country most *desirable* to wolves might always consist of such a ribbon, a seam, that is neither the rock-and-ice high country, nor civilization; but just out of arm's reach of the intolerant side of mankind. A few years ago, it came as a surprise and, I think, disappointment to some environmentalists, when a newly established wolf pack in northwest Montana began hanging out on the shoulder of a major U.S. highway, feeding on the bounty of road-killed deer that were deposited there almost daily, like sea-wrack left with tidal regularity from the fortuitous intersection of the summer tourist season and Montana's relaxed speed limit.

I think that as long as we have truly wild species, such as wolves and grizzlies, and that as long as we have truly wild country that is celebrated for its own sake, rather than for the recreational venues it can provide for us—wild country celebrated for its rankness and rawness, its forthright grace and crudity, with nothing else necessary to justify its existence, its protection—then we as a culture, a civilization, will likewise reach a purity of wildness, strength and a freshness, in our culture's collective consciousness.

But if those species and the quality of that last untouched country is lost, then we will have taken yet one more step away from the ultimate freedom that only a wild landscape can provide.

These are parallel grooves, *wolves* and *wilderness*, cut in bedrock by the same geological forces. The striations do not necessarily overlap, but parallel a similar terrain of both spirit and landscape.

In the meantime, we are carving our own striations in stone, but at cross-angles to the paths of wolves and wilderness.

Where their grooves and our striations intersect, there is dissonance, confusion, gracelessness.

But even that confusion is better than silence.

If the sheet of our force slides entirely over and across theirs, obscuring it completely, will not something vital and wonderful die within us as well?

Opponents of protecting these last little pockets of public wild-lands (which are generally managed by the U.S. Forest Service, under the jurisdiction, strangely enough, of the Department of Agriculture), will of-ten attempt to erase not only the last unprotected cores of wilderness, but even the idea of wilderness. Opponents of these places' protection will argue that wilderness has never existed, that it's a myth, and that for thousands of years—since the time of the last Ice Age, anyway—native peoples were spending every bit as much of their time and energy altering their land-scape as we do now; that hunter-gatherers were racing full-tilt through the woods with flickering blazons of pine-torch, swabbing the woods with fire, *painting* the woods with fire, achieving in that manner a vegetative manip-ulation that was conducive to producing more forage for their prey species.

Such a comparison is junk science at its worst. Even if every adult Indian in the Rocky Mountain West had been able to stop hunting and fishing, trading and warring, to commit themselves wholeheartedly to scampering through the forests with torches (imagine the forest fire scene in *Bambi*), such events would have been but a tiny fraction of the larger background phenomenon of naturally occurring fires.

A few native clans and bands and tribes, trying to survive in all seasons against and within the wilderness, could never be the equivalent of the consumptive desires of a technological nation of three hundred million (or a global population of six billion), armed with bulldozers and chain saws and feller-bunchers, carving roads into these last forests. At last count, nearly half a million miles of road had been built through our country's national forests alone. Just how many more miles do we need?

How hungry we are for these last public lands! How furiously industry fights to free up the crumbs for the continued, subsidized extraction of the public's wild treasury.

And though the argument that the Indians altered the wilderness is a specious one, it is effective; for in pausing to argue *did-too, did-not*, and to argue about the scale of comparisons, we are deflected from what should be the main force of our discussion, which is simply that some of us in this century still need and want wild country.

Even if we protected every last bit of the remaining wild country, the world would not stay the way it is now; time, and the momentum of hu-manity—the algal bloom of our success—would continue to blossom. Wildness will continue to be diminished, as our force and presence increases.

So how can we be considering *not* protecting all of the little that's left?

Even worse: how can we be failing so? No additional wilderness has been protected in northwestern Montana in twenty-five years. That statistic, more than any other, I think, delineates our twin histories of greed and timidity.

My own valley, the Yaak, is in many respects in danger of becoming an island, not quite cut off from the rest of the world, but with a significant geologic feature, the Purcell Trench, bounding it to the west, and a significant man-made feature, Lake Koocanusa, bounding it to the east. (An entire wilderness valley, the Ural Valley, which once connected the Yaak directly to the Whitefish and Glacier mountain ranges, now lies beneath a hundred feet of water; a dam was built there less than thirty years ago, constricting the Yaak yet a little tighter, making it yet a little more vulnerable . . .)

To the south, this island of a valley faces another gauntlet to the outside world, not insurmountable, but significant. A thin but formidable triple-layer of resistance separates the Yaak from the rakier icy jags of the Cabinet Mountains: the screaming, howling tracks of the Burlington-Northern Railroad, the hiss and roar of U.S. Highway 2, and the rushing wide rapids of the Kootenai River, largest tributary to the Columbia.

Only to the north—in Canada—does the Yaak continue to have full freedom to the greater world beyond, though even that source of wildness is becoming compromised by roadbuilding, clearcutting, and mining activities across the border that rival and at times exceed even our own appetites.

And within the island or peninsula of Yaak reside smaller islands, remnant, untouched distillations of the wilderness reduced, the wilderness scribed by roads and clearcuts. Thirteen little cores of wildland still exist in this valley, each 5,000 acres or larger (such a paltry sum; there are ranches in Texas a hundred times larger than these last little groves). Are they really still even worth fighting for? *Yes.*

It's amazing to consider how fragile wildness and wilderness is at this point in the cycle of the human drama.

In many instances, the species that help create the unique braid of wildness have taken their last refuge in these islands-within-an-island—these tiny park-sized roadless thickets of refugia—where they seem to be hiding, going about their lives, and awaiting the onrushing natural history that we, rather than geology, are scribing so quickly.

A visitor to the West, and especially to the Rocky Mountains, tends to look out with understandable awe at the majesty of that craggy horizon, so serrated by the last wave of glaciers that the mountains seem still gleaming from the cut of that ice. Something about a mountain—much less the view of a seemingly unending range of mountains—does something to our hearts: something that makes a person proud and excited to be alive.

But that joy is a geologic wildness, not necessarily a biological one. What are the consequences, for instance, when a vast geological wilderness—a million acres, say—is inhabited by a scarcity or even absence of the very residents (think of those residents as islands themselves: grizzly, wolf, wolverine) that help give the landscape its wild breath?

What is the result, the taste, of such a disparity? Is this not the point where a real living, breathing essence—*wilderness*—becomes instead more like the echo or a shadow of a thing?

In my valley, for instance, the populations of many of the things that help create this moving braid of wildness within the dark forests are down now to single or double digits. I think these alarming numbers illustrate Montana's vulnerability, and yet show how close hope might be for New England, as epitomized by Vermont, after nearly a century of rest.

If you discount the West's spectacular glacial rockscapes (which the Yaak's low jungles ignore well; our swamps and soft-mounded ravines and hollows are reminiscent more of the Appalachians or the Adirondacks), then it occurs to me that the two ecosystems are perhaps not as separate as has been historically imagined.

My valley, and much of western Montana, is held in federal ownership, whereas much of northern New England is held in huge tracts of so-called private ownership, if indeed you choose to believe the ruse that multinational corporations such as Plum Creek Timber, International Paper, and so forth—the dominant landowners in New England—are just like you or me, each one a private citizen, with all the rights and privileges (but none of the attendant responsibilities) thereof.

In both instances, the landscapes have endured in many places an unsustainable rate of forestry that has been referred to, even by the administrators of such programs, as a "liquidation."

Northern New England, like northern Montana, is for the most part sparsely populated by humans, barely inhabitable but to a hardy few—adrift in snow and subzero temperatures in the winter, seething with a bedevilment of insect life in the summer, and muddy as hell in spring. (Are four weeks of glorious autumn really worth all that?)

Just as the Yaak and a few other wild places like it still stand at the threshold of losing this wildness (which has been millennia in the making), northern New England stands at the threshold of regaining such threads of wildness and vigor.

Our culture now largely subsists on too many stories, I believe, taken from the long-ago—stories that we were told back when the wilderness was indeed a physical vastness, a tangible reality that was capable of matching the equally enormous metaphorical dimensions of idea, or emotion, with the story that was being told.

The wild has become not only finite and measurable, it has become tiny.

This is how fragile, how Noah's Ark–like, the situation has become in the West: Many valleys in the West have lost complete lists of species, already rendering them uninhabited by the wild ones—home now only to echoes and old stories; and even in a valley like the Yaak, attached as it is like a peninsula to Canada, we have only those twenty or so grizzlies, the one or two wolves, one occasional woodland caribou, a handful of lynx and wolverine, perhaps a dozen pair of adult bull trout, one remaining population of inland redband trout (in the headwaters of the last remaining uncut basin in the upper Yaak)—and on and on, down to the bitter end: so close to zero.

Almost as close as, say, Vermont.

All of this is to say that wildness, wilderness, is a mesh, a fabric, of the organic and the inorganic and the animate and the inanimate; that wilderness is comprised of not just the windy stony peaks of the high country, but of the vigorous braids of life moving through each landscape.

And the wolf, or wolves, are but one braid in this animate wildness.

From a biological standpoint, they perhaps are not even the wildest. Being "plastic," they are capable of stretching their wildness a bit farther toward humankind, like emissaries with an important message. (Perhaps we'll know New England has recovered its wildness almost fully when the wolverine returns, along with the caribou and lynx.)

Certainly, the wolves are at least as mysterious as they are wild. They possess their own language, their own comings-and-goings, their own force and aura; and they are both drawn to us, being canids—the rootstock of our domestic dogs—and yet fearful of us, having been persecuted across time.

We don't quite know what to think of them, and probably never will. They provide the physical landscape with some wildness, but it is in our hearts where they plumb even wilder, deeper territory.

Back to this notion of Vermont as Montana (for I think of Vermont as the quintessential New England state).

Look at a map of Montana. In the entire northwestern corner, all the way up to Glacier National Park and even over into Idaho, only one slender ribbon of land, the Cabinet Mountains—less than 100,000 acres of rock and ice, less than a mile wide at its narrowest point—in what is the wildest quarter of the wildest state in the continental U.S. has been designated wilderness.

Consider a similar chunk of country in upstate New York—the Adirondacks—or in Vermont and New Hampshire combined, or in northern Maine. These states, long considered the sissy East—tame, domesticated—actually have more designated wilderness than does northwest Montana. And not only that, New England is moving in the forward direction, protecting even more, while Montana regresses, losing eligible lands yearly. It has been twenty-five years since the one and only wilderness area in this corner of the world—northwest Montana—was protected.

An entire generation has leapfrogged across time, missing the opportunity to protect anything else for coming generations.

I keep thinking about the lines of force in the world, the vector of physical forces, such as weather and geology, overlaid on the meandering movements of time, with these twin components conspiring to create history.

In the mountains where I live, you can often find the swirls and curves of metamorphic pressures—plunges, ascensions, reversals—in geological cross sections. But on the calm surface of these formations, there's rarely any indication that such turbulence was occurring below.

Instead, you tend to find only the crudest, most direct lines of movement—grooves and furrows scratched into the stone's surface as straight as if drawn with a ruler. A glacier, either in retreat or advance, carries beneath it an infinitude of loose stones and jagged rubble that it has picked up in its travels. As the glacier slides across the landscape, these loose fragments caught between bedrock and surface ice carve and scrawl striations across the land as a man or woman might leave markings upon a page—though these markings beneath the ice do not wander or weave with the curls of our own passionate language, but seem instead to carry the message of some more patient, unalterable idea residing just beneath us.

As if the world below the surface is all coils and loops, while the surface itself is all straight lines and gridwork.

The latitudinal line that is drawn across the northern tier of our country, separating us from Canada, clearly bears in its making the mark of man, or some other mindless, relentless, unimagining thing: a line as deft and sharp as the cut made by a surgeon's scalpel.

Like so many of our boundaries, this northern border is a political line, not biological or geological. In the forest, crews have sawed a thirty-foot wide strip along this imaginary line, in a crude attempt to make the intangible tangible. It is a very strange feeling to be walking through the dense forest and happen upon this linear blaze of light, to venture across it carefully, and to then resume your walk through the dark, cool forest, with two thoughts in your head now, rather than one.

Your body, and all the available or accustomed senses, are telling you that nothing has changed, that you are still in the same ecotype, even the same watershed—but now another part of you is telling you that you have left one country and entered another. The first several times this happens, you cannot help but feel an almost dizzying disparity—entirely psychological, but dizzying nonetheless—that now you are officially an alien.

No matter how many times you may pass back and forth across that line, you will always feel this confusion to some degree: the brute territoriality of our visual marking as inescapable and unambiguous as a shout, a declaration, of possessorship.

You can feel it, every time you cross the line—the arbitrary and intangible made real, a conception, an idea metamorphosed into a physicality—and I think the wolves can feel it, too: and I think that, over time, this seemingly capricious overlay of a boundary, this thirty-foot stripe that bisects North America, a boundary sprung largely from our imagination, has accrued a spiritual and psychological presence that transcends what it once was, which was previously merely symbol. I think this physical line might be one of the many reason no wolves were known to have denned in the Lower Forty-Eight (with the exception of Glacier National Park) for over sixty years.

It was just a thirty-foot clearing, like a line in the sand, but a spirit and a force underlay its genesis—a negative, possessive one—and I think the wolves understood even without the benefit of experience that the mood, the political climate south of that line was not welcoming to them sixty years ago.

We want to believe that animals are wiser, less easily controlled or duped by our artificial constructs in the world—privy, instead, to more powerful and elemental truths than are we, such as the comings of storms, and the immutable desires for migration, nesting, and survival—and I think without question that they are; but I see no reason also that this physical acuity should prevent animals from sensing, and even understanding, the meanings of our own boundaries, borders, structures, and creations, and the impulses that led to them.

Beyond the hard-earned, acquired knowledge of Canadian wolves that they would be shot if they crossed the border, I feel certain they were also able to understand clearly the meaning of that physical border, and the consequences or risks involved in crossing it, so that for all this time that boundary has acted like a nearly impermeable barrier of spirit, to cross which any migrating wolf would have to think long and hard.

And while many biologists will tell you that wolves' return to the States, after such a long absence, was spurred by the milder winters of the 1980s, with the resultant build-up in deer herd numbers, I think it was at least as much a recent softening of attitudes on the southern side of the border that welcomed them across; or if not welcomed, then eased the resistance enough for them to begin trickling back into the country that had once been theirs.

A fireball of an activist lives a couple valleys over from me. His name is Steve Thompson, and I am awed, every time I am around him, by his knowledge and his energy.

No one loves wild country more than Steve. I consider my valley lucky that Steve knows and loves the Yaak, and is able to put in a good word for it on occasion, from his home in the Flathead. He's a flaming moderate, entrenched firmly, with bulldog tenacity, in the radical center.

As Steve believes in wilderness, believes in it fiercely, and is unafraid to say the word—quite a rarity, in northwest Montana—so too does he believe in logging, if properly done. He believes in the nobility of the profession, the inherent integrity in spending one's self so completely, either physically or mentally, and he understands that some of the forests of northwest Montana are a hundred years out of line from fire suppression and the high-grade logging of the past: choosing the best and biggest trees and leaving behind the junk—crooked trees, ill-shaped trees, and trees less resistant to the stresses of the world such as flood, fire, wind, insects, root rot.

With high-grade logging, we have basically traded away, *no*, given away —no, *paid* to have hauled away—the golden apples, leaving behind only the rotting or bitter fruit. But that bitter fruit is still wild, and we still love it; even in its bitter state, it possesses an echo of wildness.

On those injured public lands that we have so drastically mismanaged with our experiments, beliefs, and science of the past (science that is as obsolete now as our current science will be twenty years from now), Steve sees incredible opportunity for a willing workforce comprising skilled woodsmen and woodswomen to operate solely outside the boundaries of those last unprotected roadless areas, working from the already-existing gridwork of nearly half a million miles of logging roads.

"The frontier is over," says Steve, speaking of the mentality rampant in previous decades of industrial forestry on the public lands, particularly those in the slow-growing higher elevations of the arid West, on which the timber companies never intended to log on a sustainable basis in the first place.

What they practiced instead was corporate raiding, pure and simple, a liquidation of the public lands fueled by human greed. The frenzy peaked in the mid- to late-1980s, as multinational timber companies, seeking to stave off hostile takeovers by competitors who cast eyes upon their private holdings—*forests*—had their accountants measure their own companies' worth by the rawest, crudest, most unimaginative and short-sighted basis available: the net worth based on complete liquidation. Zeroing out the forests: counting every inch of timber, every marketable stick, quantifying that volume, and then converting that fiber immediately to dollars.

To keep the raiders at bay, the timber companies, which had already been overcutting in attempts to juice their quarterly profits (to make them attractive to stockholders, to raise capital, to expand, etc., etc.), now had to cut even harder and faster, converting their capital—*forests*—into cash, with which to defend against the frenzied corporate raiders.

In the Yaak, it took Champion International only a couple of years to finish off this liquidation. In peculiarly American fashion, Champion saved itself from being devoured by devouring itself; though again, in typical American fashion, the damage did not stop with the scabrous legacy of scorched-earth clearcuts, but continued to expand, wreaking havoc like some mythic Hydra. In cashing out, Champion had plumed an incredibly unnatural volume of timber through the local sawmill. Employment, as a result, was at an all-time high. The rural counties had never had it so good; or so it seemed.

In actuality, the rural communities were being held hostage to fortune. Champion, and other companies like it throughout the West, went to the county commissioners, and to the Forest Service, and presented them with what was essentially a legal kind of extortion. They told the Forest Service that timber companies would now need to turn to the natural forests for their timber supply, and that they would need to keep up that same rate of liquidation, or else the timber companies would have to close down the mills and leave town, heading back south to Mississippi, Georgia, and Arkansas, where the trees grow faster, if not stronger.

The timber companies went to Congress, too, because that's who controls the Forest Service's purse strings, just as the timber companies control Congress' purse strings . . .

Well, you get the picture. The timber companies won, while a list of the losers included wilderness, rural communities, independent loggers and small sawmills, wolves, grizzlies, elk, peace, quiet, solitude, clear water, trout . . .

Under this insurmountable pressure, both the Forest Service and the U.S. Fish and Wildlife Service folded, allowing the timber companies to cut at record levels on the national forests. The timber companies achieved record profits, then invested the money in labor-saving machines and technology, and paid out increased stockholders' dividends, rather than using those profits to increase jobs in the community or to raise worker's wages. Finally, when Congress could no longer break the laws nor rewrite them—when the cutting finally began to slow down—Champion did what they had planned to do all along, which was to flee to the South, in a classic and utterly shameless display of cut-and-run, leaving abandoned mills and a legacy of hateful name-calling in their wake.

Except they weren't quite through. On their way out of town, after having clearcut their private lands along the river bottoms, and having clearcut the public lands in the steep hills and mountains of the national forests, they now converted their limited partnership corporation (a crafty device by which, unlike the rest of us, they had never had to pay taxes on any of their earnings beyond a certain level) into a real estate company.

They divided those massive clearcuts into subdivisions that were already prepared for home construction, what with all those pesky trees cleared away—and rather than abandoning their cut-over lands to repair in peace, they ignited a new land rush, opening up the riparian areas to a glut of ranchettes whose negative effects exceeded that of almost any clearcut.

They couldn't have hurt the ecology of the valley more if that had been their sole intent.

A warrior like Steve Thompson doesn't give up. In addition to lobbying for prohibitive retroactive property tax increases on timber companies who convert to real estate companies after denuding their lands, Steve has been involved for several years with helping establish wood product certification standards for so-called "smart wood" or "green wood"—the logger's equivalent of organic gardening, in which timber that has been harvested in a sustainable, even benign, manner, is given a seal of approval.

With savvy marketing, particularly with the use of the Internet, "smart wood" can be a means whereby mill owners or loggers can obtain the highest possible price for their products. An economic motivation beyond pride of craft will help ensure that the forests are treated the way loud residents want "their" forests to be treated. As well, the certification can help mill owners gain market share, which economists say is at least as important as price.

I wouldn't know. I just know it sounds like an exciting idea. And while I'm enthusing about the abstractions of it, Steve is doubtless at this very moment in some godawful boardroom, talking to loggers, mill owners, county commissioners, economists, silviculturalists, and environmentalists, trying to hammer out the specific protocol for what makes a piece of wood "smart" or "not smart."

I cannot tell you how far that world is from me. Just the thought of it makes me want to crawl into an old-growth grove of cedars, hemlock, and ancient white pines, and cease thinking.

To lean against one of the giants and simply rest, and let Steve do it all —to let Steve keep doing it all.

I have been thinking for the last few months, off and on, about glaciers. I am burned out, I think, on how slow change has been in coming to northwestern Montana, regarding the preservation of the last few remaining roadless areas as wilderness, and at how resistant we've been as a culture to reforming forestry, the dominant land-altering management practice in this region.

A glacier isn't just a chunk of ice, static, frozen in time. We know that a glacier is like a river, sliding through time, up one valley and down the next, with the laminar grace of wind or flowing water. It is only when we try to compress the life of a glacier into the scale of our own short lives that the life leaves it, and it ceasees—or so it appears, to our blind eyes—to move.

Glaciers are always moving. Who dares to say that they are not: that because we cannot see a thing happening, it is not happening?

The common definition of a glacier is any patch of earth where snow remains present year-round, with a depth of ice accumulating across time. What interests me is the point where a glacier, having invested decades in patient endurance, finally gains enough overburden and ceases to act like a pile of snow, and metamorphoses into a slippery, living fluid, and begins flowing down the mountain: carving a new world, finally, and setting new rules.

I cannot help but think that efforts to protect the last roadless cores of public land is exactly like the formation of a glacier; and this is a thought that gives me solace. Zero acres have been protected in the Yaak Valley, nothing has been achieved, since the national forest system was created up in this corner of the world—almost a hundred years without wilderness— yet perhaps one day our effort will accumulate enough mass.

I used to have a different vision with regard to environmental activism in this part of the world. I used to think that up here, it consisted of the tedious, enduring stonework of a mason; that the activist's duty in such a situation—outnumbered a thousand to one—was to stay calm and to be enduring; that even the greatest and most skilled artisan could never be anything more, from a historical perspective, than a laborer, laying down stepping stones across some river across which those who would come later might pass.

What I think now is that it is not like laying stones across a river, nor stacking one's stones, one's words and deeds, in an ongoing, eternal rock wall. Rather, it is like being snowed upon in winter. One's life is not nearly so significant or dramatic as a rock wall, but is instead far more silent, made up often of a lifetime of snow. I have begun to think that the conceit of our lives, our hearts, as hot furious maelstroms of passion is but a romantic fancy; what we really are is nothing more than windblown snow, beautiful, yet ephemeral when measured flake by flake.

We would like to think of ourselves as slabs of mountain: square-cut, durable, and imminently useful. I think however that the opposition—those who would kill and eat the wilderness—are the stoneweight, and that we and our works, however passionate, are but snow, melting unto vanishing in poor years, though in good years beginning to accumulate in skiffs and ridges, drifts and mounds.

For a long time, nothing happens. It's just snow: silent, beautiful, monochromatic, relentless, heartless—or so it seems. Thirty feet might fall in the mountains, one winter; by summer's end, twenty-nine feet might have melted.

But then winter—the time of beauty, the time of work—comes back around again. Thirty feet might fall again, and once more, beneath summer's broil, twenty-nine more feet might vanish.

Nothing is happening yet. It's just a bunch of old crusty snow and ice. But something is getting ready to happen.

Our work—no matter how seemingly unproductive—is not for nothing.

Depending on slope and aspect and density—usually a function of moisture content—it takes between 165 and 200 feet—50 to 60 meters—of ice to make a glacier come to life.

Once that proper load weight is established, the glacier begins to move, begins to flow continuously, behaving now as a plastic or a liquid, rather than as the brittle solid it had heretofore been, trapped and frozen for so long.

I've often wondered what it must feel like to the mountain, when the long-building glacier first begins to move. What do the inhabitants on that mountain think?

Surely there must be an audible, tangible release of relief, in that interface between glacier and mountain, when the glacier does begin to move. What must it feel like for the glacier, as it finally comes to life, a blossoming or flowering of ice?

Some of the ice at the bottom of certain glaciers has been measured to be 25,000 years old.

I do hope we get the last roadless cores on the public wildlands protected—permanently protected—before that kind of time, but if not, we will keep doing what we do, will keep saying what needs saying, even if for all intents and purposes it seems as if we are speaking in silence, season after season, with nothing happening.

For several generations in northwestern Montana, representatives of the extractive industries have been pushing the fear-button, prophesying that if roadless lands in the natural forests are designated as wilderness, then the private land owners next to those national forests will be evicted.

Hardly anybody will say the w-word. We talk about bulldozers, talk about hunting and fishing, talk about the basketball playoffs; but then sooner or later I'll mention the w-word, just to keep the idea from being forgotten, and everyone will get all upset for half a year or so. Threats get made, names called.

My neighbors try to make it easy for me. I've had conversations of varying lengths (some very short, it's true) with almost all 150 of them, at one point or another in my fourteen years in the valley, and I don't think it's inaccurate to say that roughly 99 percent of them would agree that they would like to see the valley stay just the way it is, and keep the roadless areas roadless. But why call it wilderness? Why not just leave it alone?

Because I'm tired of fighting, for one thing. When I moved up here, I used to be a fiction writer. I loved that craft, that calling. I've had to all but abandon it, to speak out instead for another thing I love now just as much as language: the woods. These woods. And I need it to be designated as wilderness because, frankly, there are but a few of us up here paying the dues, fighting back sale after sale in one unprotected roadless area after another. By and large, not a lot of marketable or accessible timber is left in these last roadless areas—which is why they never had roads built to them in the first place—but all I can surmise from recent activities is that industry, and Congress, want it all.

In the last five years, at least seven timber sale proposals have been made in the Yaak's last roadless areas; each time, public opinion has been able to turn them aside, for a little while longer.

Most residents of the valley aren't aware of these stories. They look out their windows every morning, or drive up and down the Yaak road, admiring the view, and they tend to their gardens and live quiet lives and wonder, I know, *What's all the fuss? Why does this Bass guy like to fight so much? Can't he see that things are better, that logging practices are improving, and that for the most part the Forest Service is leaving those roadless areas alone?*

My neighbors try to make it easy for me, but of late I get the feeling that they are growing weary with my inability to take the proffered chance, the palm branch. What would you think about calling it a natural area, instead of wilderness, they ask—or how about "heritage forest," or "roadless core," or "primitive area," or even "national roadless monument?" What about "National Recreation Area?" How about an eighteen-month moratorium? How about a ten-year cease-fire? What about other kinds of new designations?

To all of these I have to say no, it's not enough. I have to stay firm and stand my ground, demanding wilderness, which is the only complete and permanent protection a forest, or a watershed, or a mountain, can obtain. All the other designations carry various loopholes: for helicopter logging, snowmobiles, motorcycles, gold and silver and copper mines, hydroelectricity projects . . .

I'm not even saying that I'm against those activities, nor even that I'm opposed to them being executed on the national forests *in certain places.* All I'm saying is that I want to know—need to know—that these last little roadless islands will be protected, permanently. I do not want to have to travel down to Troy, or Libby, or Helena, or Missoula, or Washington, D.C., week after week, year after year, to keep trying to turn back the road-building from those last little refuges. I want my life back. The frustrating irony is that most of my neighbors want to offer me my life back, if only I will discard my beliefs, my needs. If only I will think of myself, and the present, and sell out the future, which is so suspect in any event.

Saying that you want or need wilderness in northwestern Montana is not quite the same thing as saying you prefer Chevrolets to Toyotas, or that you once voted a straight Democratic ticket, or that as a Baptist you think the Methodists are all a bunch of hooey. In the woods, and in the country, where the allegiance of almost all the residents, save for a few real estate developers, lies so intensely with the land itself, we are talking about nothing less than how to manage another person's soul.

I have been thinking about glaciers, and have been taking solace in their grace. As long as there is ice, and snowfall, there is hope. Once that adequate load or pressure has been achieved—50 to 60 meters—and the ice begins to flow, a second kind of movement occurs, at even greater depth. In addition to the flowing of all the ice resting beneath that 60 meters, the entire glacier itself begins to slip from its contact with bare earth, with bedrock, sliding now like a wet tennis shoe on ice. Sliding within sliding, sliding upon sliding. You can imagine, once this phenomenon is achieved, how very hard it would be to stop it. You can imagine how the glacier would begin with claws of ice to write its words, its sentences, its stories in stone upon the previously implacable mountain. You can imagine how such etchings might finally begin after thousands of years of waiting to be released—to get the mountain's attention, and that of any other observer who might happen to be around.

These glaciers—lying so patient upon the stony, intractable mountain, will, in the end, have their say almost completely. They will carve and shape and scour all the way down to the core, the bottom, cutting the mountain itself: the very thing that birthed the glaciers by creating those weather systems that gave rise to the snow and clouds.

It all started out as only an inch of snow, two inches of snow, upon a great mountain range. Just a skiff at first. But the steady patient snow can and does change the stone.

Whether you want wolves or wilderness, or both—it helps, sometimes, I think, to just stand for a moment and stare up at the snowy tops of the mountains.

Just as exciting, to one whose heart is increasingly emboldened by the patience and strength of glaciers, is the unexplained, unpredictable phenomenon of glacial surges, in which the slow advance of some glaciers is characterized by periods of extremely rapid movement. A glacier might be moving along in its normal manner, only to speed up for a relatively short time, before returning to its normal rate again. The flow rates during surges can be as much as one hundred times the normal rate. No one really knows why such surges occur.

The thing I most need to believe, even when there is no evident cause to do so, is that even when it seems that things are not happening, things *are* happening—at the surface, and especially below the surface. Complex, subtle things.

There is nothing about glaciers that does not give me hope. Even more exciting than the graceful, riverine movement of things that appear to be immobile chunks of ice jam, is the miraculous news that even with glaciers that are having a rough time, with the ice front retreating, the old durable ice *within* the glacier will often still be advancing, like a rip tide running in reverse, so that once that momentum has begun it is as if the glacier is its own living entity, its own organism, intent upon and committed now to a destiny that seems more biologic than geologic: one that seems to radiate a force of will.

And surely, seen from some great distance of either space or time—

from a hundred miles up, or across the span of ten thousand years—the movement of the glacier would seem like a sinuous galloping thing, sliding across the landscape with an elegant gait, and leaving behind tracks etched in stone, and spoor the size of small villages.

I believe deeply in Wendell Berry's advice that, "To enlarge the areas protected from use without at the same time enlarging the areas of good use is a mistake." In an attempt to demonstrate to my neighbors that I care for them and their ways of life, as much as I do for the wilderness—believing as I do that like me, they too are of the wilderness—I put aside my wilderness advocacy for a while and found myself dedicating the autumn and winter of last year and this spring to work with a small local group to lobby hard in Washington and Missoula and Helena, in Troy and Yaak and Libby, to secure one of the nine pilot forestry programs authorized by Congress in Montana, Idaho, and the Dakotas.

In these pilot areas, a new style of logging would be practiced. Rather than the old paradigm of focusing on how many board feet a unit could yield in a single cutting, two new objectives would be paramount: the desired "end result" of the forest, and community participation in the planning process.

Loggers would be awarded contracts based on the quality of work they could perform, rather than the amount of timber they could scalp in the shortest amount of time. As well, the program would favor the employment of local workers who, working on their own home ground, would have more motivation to take the weaker, lesser trees, and leave the strongest and the best. The emphasis of sites selected for this program would be on the restoration of damaged landscapes and watersheds, rather than the further invasion and liquidation of healthy ones.

After six months of hard work, our little pro-roadless group in the Yaak was able to secure a much-coveted position in this federal program. However, my wilderness views—beliefs, needs—made too many people in the community uncomfortable, even distrustful, and many of them asked—some of them demanded—that in order for the project to go forward, I not participate in it. (Further disheartening was the fact that the Forest Service had planned a 200-acre cutting unit next to my property; I'd hoped it could be an experimental unit in this project, a unit that would find a treatment that differed from the Forest Service's prescription, *clearcut*.

I had very badly wanted to be a part of the pilot project. I believe in the theory, the process, even now. I wanted to walk through the woods with my neighbors, saying, *Let's take this tree. Let's leave this one.* Like them, I wanted to have a voice.

I still believe that there is opportunity for sustainable forestry in the Yaak. The valley has been turned upside down: It was once 50 percent old growth, but in the last half-century rampant clearcutting and taking of the best trees instead of the weakest has converted the valley to a young forest of overstocked, crowded, weaker trees; nearly two-thirds of Western forests now consist of trees less than 17 inches dbh (diameter at breast height). In many of these post-harvest units, opportunity exists for the restorative work of thinning, both commercially and precommercially.

Even after I brought this new logging project to the valley to put local people to work, opponents of my needs—opponents of wilderness—continue to spread the rumors that, because I favor wilderness, I am against all logging, and furthermore, that I want everyone—myself included—evicted from the valley.

Wearier, if not smarter, I have retreated to the far perimeters of the community, for now. The place where everyone wants me, the place where perhaps even the landscape wants me, and hell, perhaps even the place where, when all is said and done, I myself want to be—though it does not feel that way to me. I continue to wish to have this issue settled, and to become a part of my community again, rather than apart from it. I can envision some day, a very long time from now, when we're all healed and rested, trying something cockamamie again on behalf of that quaint logging culture that I continue to support because I believe in the principle, and because they are my neighbors, but which, truth be told, gets a little less quaint with each passing year.

I need only to consider the work of certain Yaak loggers, however, to remember that this vision of responsible logging is not a pipe dream, it is a reality. We need only to create authority and economic incentive to reward their kind of work, rather than the other kind. Even those who hate me worse perhaps than they hate anyone else in the world are entirely capable and desirous of doing stellar work, given the opportunity.

Maybe it's part of the glacier's process to lose some of its fire. Or for the nature of the fire to change, as if metamorphosed, under the pressure of time and distance traveled, or not traveled. So much labor, perhaps

the better part of a life's labor, with not one single gram of tangible difference. It is not a giving up or even a wearing thin, but instead, a reconstitut-
ing of desire. In some cases, fire is too quick an instrument for the work
that needs to be done. In some instances, ice is the tool that is called for—
the ice of decades, or the ice of centuries. If you cannot, even in your best
efforts, attain speed, then perhaps you can gain solace from images and
models of power: the great sheets of ice that sculpted many of the wild
places we find ourselves fighting for. The great masses of snow and ice that
scribed *their* direction, no one else's, into every ravine and peak and cirque
of these stony landscapes, and that laid to level the resistant landforms,
entire mountain ranges of obstacles. Retreating, advancing, retreating,
advancing . . .

In the static or even backward-moving attempts of my goals, I have been
taking solace in the patterns and history of glaciers. It is a pleasure to know
of the nature of crevasses—cracks appearing in the zones of tension that
are created as the glacier creeps across irregular terrain. Though the jagged
fracture of these gaping cracks may reach as deep as 50 meters, for the most
part the crevasses are merely surficial, even cosmetic expressions of the
struggle below, and as such, insignificant in the flow of things. At depths
down below that 60-meter mark, the plastic flow of the glacier remains relatively unperturbed; seamless, with the hidden and earnest momentum of
its mass and, I like to believe, its desire.

You think things are moving very slowly, or even not at all. But you
must not forget those shiftings and releases of built-up, buried hydrostatic
pressures, little-understood, wherein one day the glacier surges, suddenly
like a skater free and clear, traveling as much as 180 feet in a day. (Beneath
the glacier, at its base, are runnels of gushing water, creeks and rivers honey-
combing through the ice in myriad and unmappable, unknowable braids,
like the convolutions of a brain's folds. It's theorized that, like the miracle
of lock-and-key, those shifting braids of river tunnels beneath the ice one
day click into such a combination as to provide an immense hydraulic lift to
the project, and the whole glacier is fairly catapulted down the mountain.)

I take solace in the agony of the glaciers, and their sloppy detritus, their
tortured yet elegant residue. A glacier rasps, roars, grinds, hisses, gushes,
and spurts as it moans and hollers its way down and through the mountains. It spreads before it, like the rimed breath of some advancing colos-
sus, the clattering terminus of boulders deposited as till and outwash.
Sometimes this breath, this terminus, is carried vast distances, and once

deposited in a new land, these deposits can play a significant role in transforming the physical landscape.

A glacier leaves behind a legacy of troughs and drumlins, arêtes and horns and tarns, cirques and kames and eskers, kettles and moraines. A glacier is transformed daily, even while it seems to represent the very heart of consistency. It adapts, shifts, flexes, readjusts itself like an animal bedded down in a straw nest, turning round and round, seeking the right and comfortable position. But always moving. A dry paste is often deposited across the land by glaciers, the grist of all that work, all that grinding, which is called rock flour; it exists briefly, where it separates from the rivers draining from beneath the melting glaciers, then is whirled away by the wind and cast across the globe, having been visible, measurable, for only a moment.

We don't even know where they come from, really. One theory that was once prominent, then fell out of favor, but that has been embraced again is that glaciers are the result of a wobble, a hitch, in the earth's rotation. An imperfection, if you want to think of it that way, but what a beautiful imperfection. A Yugoslavian scientist named Milutin Milankovitch theorized that the climatic changes that give rise to the build-up of glaciers occur as the result of slightly cock-eyed variations in the shape, or eccentricity, of the earth's orbit around the sun—as if, in some great and vast gearworks, a tooth in a cog were chipped or missing—as well as changes in *obliquity*, which is the difference between the angle of the earth's axis, and the plane of the earth's orbit, and, additionally, from the simple, charmed, wobbling in the earth's axis, like the spinning of a water-balloon, which is called *precession*.

These factors make the earth a moving target for receiving the sun's radiation—sometimes we receive more, other times less—as if dodging some of it, in our wobble. The glaciers get built, or not built, simply, miraculously, because the earth is canting one one-trillionth of a degree in *this* direction for a long period of time, rather than in *that* direction.

When I am alone in the woods, and the struggle seems insignificant or futile, or when I am in a public meeting and am being kicked all over the place, I tell myself that these *little* things matter—and I believe that they do. I believe that even if your heart leans just a few degrees to the left or the right of center, that with enough resolve, which can substitute for mass, and enough time, a wobble will one day begin, and the ice will begin to form, where for a long time previous there might have been none.

Keep it up for a lifetime or two or three, and then one day the ice will begin to slide—it *must*.

I do not mean to dismiss our fiery hearts. I mean only to remind us all that our lives, our values, are a constant struggle that will never end, and in which there can never be clear "victory," only daily challenge; and that after the glacier passes over, all it will leave behind will be sentences and stories written in stone, testimony not to winning or losing but rather, simply to what path the glacier took, while we were here, and whether we were the mountain or the ice.

Isn't a lot of this just the indulgent story of my angst?

I don't think so. I know it appears that way. I think many of our environmental battles and concerns are converging—separated less in time and distance than ever before. It's eerie to see that Plum Creek Timber has leap-frogged into northern Vermont, and is pursuing their same scalp-and-subdivide program there. It's eerie to see that wolves just across the border there: want, it seems, to drift back into their ancestral ranges, but get shot or trapped every time they cross that line.

It's strange to see sustainable, community forestry initiatives popping up in northern Vermont, though Vermont's programs seem fifty or even a hundred years ahead of Montana's, and seem to be proceeding with smooth sailing. Again, I have to remember glaciers.

It's strange to see that both states are fighting for the preservation of their last remaining wildlands.

Vermont, however, had to cut all its timber and then wait a hundred years for it to grow back, in order to reclaim a second- or third-growth wilderness, and one that is—still, if even only for a little while—void of wolves. Such a realization can suddenly make the thirty-five years we've been battling for wilderness in Montana seem like a tea-party, a warm-up exercise.

Is that northwest Montana's destiny? Must we follow that same groove —see the last of our roadless areas cut and altered, the last of our wolves and grizzlies erased—and only then, a few hundred years later (trees, and forests, grow so much more slowly in the West), begin to learn what Vermont has finally learned: that wilderness is a resource too, and that with imagination and commitment, and a supple, healthy landscape, it's still possible in places to have wilderness, wolves, and working hunter-gatherer communities, if not industrial forestry?

Held tight in Vermont's stone heart, and the stone heart of the Northeast, is the lithic memory of mountains that were once as fierce and jagged as Montana's are today. In *The Story of Vermont*, Christopher Klyza and Stephen Trombulak describe lyrically the tectonic processes that give birth to the state. One great continent, called proto-North America, floating on a belly of the earth's fire, a deep-seated lake of fire, drifted, unmoored, into another land mass called the Taconic Island Arc. The two landforms merged into one newer, larger one, with intense buckling and transformation—mountain-building—the result.

"The amount of compression caused by the collision was enormous," write Klyza and Trombulak, "and faulting resulted in extensive overlapping and jumbling of blocks." Valleys, rivers, and lakes were created along the faults and rifts, and large masses of magma hissed and belched their way through fractured zones of weakness, forming granite domes and seams of mineralization, which appear even today like heaven's glittering streets of gold.

The mountains rose higher—they must have been magnificent to behold—but were then worn down by the forces of the world. The glaciers, certainly, were attracted to these new mountaintops. Then, from the eroded rubble of these magnificent, eroded ghost-mountains, a newer, softer mountain range—the Appalachians—was formed when activity occurred above a "very small, localized hot spot under the land to the west of Vermont, [which] caused the overlying rock to expand and bulge upward."

In this manner, those old buried or worn-down mountains are rising yet again, like some titanic swimmer surfacing for air. Perhaps one day again the mountains of Vermont, if they keep rising—as they keep rising—will capture new glaciers, and new stories, new landscapes, and new natural histories—will be carved by that splendid ice.

But for now the glaciers in Vermont are gone. They have been gone for only 12,500 years, but before they left, they lay across Vermont with a thickness that was more than a mile deep: dense blue ice capping the earth's fires, the earth's plans, so far below.

Whether Montana is only a hundred years behind Vermont, culturally, or twelve thousand years, geologically, makes no matter. All things will rise, all things will be worn down, before being lifted up again. If only one story could be told in the world, surely it would be this one.

The wolves were once in Vermont and the Northeast; they will be back.

In some ways, this drift toward sameness—this revolution of cycles, this rise-and-fall—terrifies me. Though part of me longs for Montana to catch up with Vermont—with the sustainable community forestry initiatives, and the value-added wood-products innovations, and the greater enthusiasms for wilderness proposals—in my own core, my own bedrock, I must confess that I do not want Montana to become Vermont: or not until a very, very long time from now.

Again, for this fear, raw wilderness—as much of it as we can get—is an antidote.

Wilderness tends to put on hold the homogenous schemes of human-kind. Wilderness tends to preserve Montana as Montana, and Vermont as Vermont. Even as all humans and all cultures become ever more the same, at least we can—with foresight and action—preserve and protect the wild places whose great riches are due in part, perhaps, to the fact that they are still unique, different from one another, rather than maddeningly, frighteningly, similar.

Why am I spending so much time talking about geology, and logging, and forestry, instead of wolves? Why not just drop the wolves into the Northeast's forests, wish them well, fine or jail anyone who jacks with them, and glide on into history?

Surely after my latest drubbing, and the ongoing, continued community shunning, I find the noble independent logger, and the logging culture, a little less quaint than before? Surely, I've reached the point of belief now that the whole mess of them, all loggers, are pigheaded bullies, puppets of big industry's rhetoric, and that if they are not foresighted enough to adapt to more innovative, practices (even as the big mills are selling them down the river), then they deserve what's coming, which is extinction?

There's a small mean part of me, a little sliver of a wedge, that thinks that way, again and again—every time I hear or read something where one or another of the industry's p.r. representatives has engaged in name-calling, badmouthing environmentalists; every time I get knocked down, there's that part of me that wants to say, *Fine, dipshit, have it your way; I'm only trying to help us stand together—have it your way.* (I think my most recent surge of truly incandescent fury was last April 15th, when I took off work to go down to town to get my income tax returns postmarked. I'd worked my ass off, had written and published five books in the last eighteen months, and as a result the Tax Man had taken a hugely disproportionate bite: everything I had left, and then some. It seemed to me an obscene amount of money to be sending back—a couple of the largest checks I'd ever written in my life—and while I was in town, motoring down the road in my old piss-ant truck that I'd bought reconditioned, vintage Jimmy-Carter, I passed some big gas-burning hog of a new truck, the kind I'll never be able to afford, that had a bumper sticker that asked, "Are You An Environmentalist Or Do You Work For A Living?" Any other day, I might have been able to laugh, but that particular day in mid-April, I about blew a clot.)

What keeps me going is not any spirited defense of the self-pitying, pissed-off dying remnants of the company-town mentality—the old boys who were told by the big companies to clearcut entire watersheds, and who did so to put bread on the table, and who now find themselves in the unenviable position of having to defend those past actions.

Instead, I think about the dying breed of loggers up here, many of whom are my friends and some of whom have logged on my own over-stocked young and emerging forest, for whom work is a kind of prayer, for whom being in the woods is a kind of sacrament, and whose pride of craft, whose signature of integrity, is not the work they take away with them on the backs of the trucks, but that which they leave in the forest uncut, anchoring that wild forest. Their names might mean nothing to the world beyond the little communities in which they live, but it is these men and men like them whom I think of when I am tempted to just turn my back and devote 100 percent of my time to lobbying for these last wild and rare, fragile and unprotected roadless areas of the national forests, rather than dividing the time I have left in the pursuit of sustainable forestry. I hope I am not making the wrong decision.

Probably what annoys me more than the old-school logging dinosaurs are the people who agree that we should keep the logging local and sustainable, and also protect the wild places—but who remain silent. By failing to speak up for peace in this generational war, they are with their passivity condoning the ongoing war; with their silence, they help keep that wall thrown up so that there *are* two sides, rather than a truer model of community, in which—as with a healthy, diverse forest—all participants are linked inextricably, unable to be shut off or unaccountable to one another.

My challenge to these silent, passive believers: Please speak your heart's truth—if not in public, then at least to friends and family. Speak of your belief that there is still room for wilderness, still a need for wilderness, more than ever, and that wilderness need not be the enemy nor the stumbling block for preserving the tattered remains of a bruised and battered, vanishing culture of logging in the Pacific Northwest and the Rocky Mountain West.

It is the silence of the passive moderates in the community, not the rhetoric of the extremes, that is dooming us to play out, again and again, these old troubled cycles of boom and bust, in which the wilderness is carved away *and* the good jobs in the woods are sucked away by the acceleration of history.

If any wild place is left—any roadless area, any vibrant core of wild, untouched forest—we need to protect it, if for no other reason than as a last example of how the world existed before we set our machines out across it in search of short-term profit.

I am not speaking of compromise, in which some percentage of these last little islands of integrity, these last little shreds of remnant wildness, are sacrificed. I am speaking only of each of us reaching out a hand to the other and asking how can we help each other. To find areas of common ground, rather than common conflict.

No roadless area can be entered without conflict. We must set them aside, out of the fight's way, if we are to successfully move forward.

What this talk of logging and forestry has to do with wolves, I think, is that the argument over forest management seems to provide an almost identical overlay for the arguments concerning wolves.

There is the hands-off "zero-cut" approach in logging, in which no trees are cut commercially from the national forests, and there is perhaps its equivalent, "wild wolf" advocacy, whose proponents believe that none of the wolves should be managed. According to these advocates, the returning wolves should not be relocated from Canada by helicopter or any other means, nor should they be live-trapped or fitted with radiotelemetry collars.

There is a moderate camp of logging, in which room is found for some trees to be cut in certain fashion in certain places in the national forests, and there is also a moderate camp of wolf recovery, in which proponents believe there is room for some wolves—not all—to be fitted with the radio collars, at least during the initial phase of recovery.

And most certainly, in logging, there is the hardline extreme, the right-wing absolutists who would not protect any of the last remaining unprotected roadless areas in the national forests as wilderness; and on the wolf issue, there are the way-right absolutists who would likewise prohibit any and all wolf recovery: that which might occur "naturally," as well as that which might come by the hand of humans.

The arguments will be similar in both sets of models, as will the tenor and passion, the showboating and betrayals by politicians beholden to special interest groups. In the end, it will be up to the people to subdue their leaders—to chase them to ground—and impose the people's will, not that of the corporations, to bring about change.

The wilderness—abstract, haunting, unnameable—will perhaps always be an enigma to the American public and American politics, and will continually unravel, as fewer and fewer people are exposed to it, and hence find little reason to battle on its behalf.

Many days, on my rambles through these last little islands of unprotected forests, I can envision wilderness designation *never* being achieved for these last little cores—not in my lifetime or anyone else's—so awful and paid-for has the political landscape become; and I want you to know, this is not a good feeling.

The wolves, however, are coming. Unlike the wilderness, wolves are not abstract to the American public. This is due in large part to the intense commodification of wolves in the world of ad agencies, but due also to that bedrock gene of affinity we have for all the large carnivores like ourselves—particularly the canids. I do not mean to underestimate the hard work that lies ahead for the wolf activists—on the contrary, the struggle will take every bit of energy and ingenuity they have got—but the wolves are glamorous and, to some degree, can be mapped and charted and quantified. That is, wolves can be commodified, and when America goes shopping, America gets what it wants. The wolves have come, or been brought, to Montana and Idaho and Wyoming, to Washington and Mississippi and Arizona and New Mexico; to the Carolinas and Kentucky, to Minnesota, Michigan, and Wisconsin, and they will be coming to New York, Vermont, and Maine, too. I believe that is a certainty, as sure as if written already in stone by the script of glaciers.

Perhaps another reason it might be fitting to talk so much about the politics of forests, rather than of wolves, is that the forest is the system by which the wolves integrate themselves into the world. I think I would be remiss to speak only of the wolf's howl and reproductive cycles and behavioral characteristics, while ignoring the larger issue, which is, *Where are they going to live?*

To focus on anything other than the forests, and the human communities within and at the peripheries of those forests, would run the risk, I believe, of turning the entire operation into nothing more than a semi-passionate shopping expedition. It's one thing to capture and relocate wolves into a new setting, but quite another to keep them happy there, and to tolerate not just their beauty, but their ecological extravagances.

As well, it seems fitting to spend so much time looking at the woods, rather than directly at the wolves, because that's pretty much how it's going to be once they return. Their presence in the woods will often be like the essence of faith. You'll know they are there, but you will rarely see them. You might see or hear a hint of their presence now and again; a cracked deer bone lying on the forest floor; a faint print in the melting snow. But for the most part, it will always be someone else who heard or saw them two or three valleys over, two or three days ago. You might see the real thing only once or twice a year, or maybe never. You might stare out at the dark wall of the forest and see only trees.

While I'm living knee-deep in wolf country, lobbying for wilderness designation for the last little scraggly wild cores of roadless land patchworked here and there on the national forests in Montana, and providing support for community forestry pilot projects—tame little visions *reeking* of the predictable and the orthodox—over on the other side of the country, activists are making innovative, impatient proposals as bold and wild as Montana's are modest and conventional.

In Montana and the West, opponents of wilderness want to take lands away from the federal government, while in Vermont, and the rest of the northeast, folks want to buy it (from willing sellers only) and *give* it, or sell it, back to the government.

Advertised as "America's Next Great National Park," the proposed Maine Woods National Park project seeks to purchase a vast acreage of industrial timberlands in northern Maine and deed that land to the federal government, which would use the land to create a 3.2 million-acre national park. In many respects, it's the exact political opposite of the story going on in northwest Montana, even though the ecological subtext is identical.

In northwestern Montana, the cutover, hacked-up lands belong to the government in the first place; you would think our task of establishing some protection for those last little unhacked vestiges would be easy, compared to the national park idea in Maine, where the dynamics of ownership are the exact opposite. There, the bulk of the ecosystem lies in private, corporate ownership. But it's a strange world. In these late, last days of wildness, it seems as if it might be easier to save corporate wildlands than those owned by the public at large and managed by the U.S. Forest Service.

Not that a national park in the Northern Forest is necessarily a new idea, for all its boldness. Over a hundred and fifty years ago, Henry Thoreau envisioned the creation of what he called a "national preserve" in the heart of the Maine woods. And the idea has been kicked around on and off since then, like the coals in a low campfire glimmering whenever a breeze stirs across them. Why, after a century and a half, does the concept seem to be moving now with such alacrity? Perhaps because no need was perceived to exist before then. Or rather, people like Thoreau and other visionaries saw the need, but no one else did, back then. (In his 1916 *Journal of Maine History*, John F. Sprague wrote: "If a portion of Maine's northern wilderness could be set apart for . . . all of the animals, and all of the song birds . . . what a wonderful national park it would be.")

Despite a lack of formal protection, however, the wild Maine woods remained wild. They were just too far away, too dense, too buggy, to be worth fooling with from a commodity standpoint. Other areas were closer to the teeth of the sawmills.

As well, the land owners in the region took pretty good care of their lands, often working in the value-added, low-volume tradition of stewardship. The economic incentive lay in sustaining or even improving the land, not in liquidating it.

All that has changed recently. More eerie coincidence in this conceit of New England's and Montana's histories running in parallel grooves, separated only by time, not distance: the Montana fugitives, Plum Creek Timber and Champion International (formerly Champion Timber, now Champion Realty) have leapfrogged from Montana to the Northeast; they presently are active in Maine, selling off their timberlands.

RESTORE: The North Woods, a nonprofit organization dedicated to the restoration of the north woods, writes of this recent change:

> Today, control of most of the region has become concentrated in the hands of a few large corporations. Driven by global pressures to maximize short-term profits, they have been clearcutting the forest, spraying toxic pesticides,

building extensive logging road networks, and subdividing pristine shore-
lands. To cut costs, they have sought tax breaks and eliminated thousands of
woods and mill jobs. Without protective action the Maine Woods—and a
valued way of life for Mainers—may soon be lost.

In March of 1999, I met with the program coordinator of RESTORE,
Kristin DeBoer, and new executive director, Michael Kellett, in Concord,
Massachusetts.

DeBoer's and Kellett's group is a cutting-edge model in high-tech press
and communications: the marketing of an idea, the creation, sustenance,
and nurturing of a concept, a dream. It's a fascinating example of the transformation process that's not unlike that of sunlight-to-wolf: the seemingly
abstract to the more specific.

I have no doubt that RESTORE will be successful. How strange it is to
consider that words on paper, and thoughts and desires, can be enough to
help summon a creature across a border.

Perhaps RESTORE's single most inspired move thus far has been the
creation and distribution of a visitors' guide to Maine Woods National
Park, designed in exactly the same fashion as all those other bright and colorful fold-out brochures that the National Park Service distributes from
the voluntary-donation drop-box when one enters any of the other national parks. The brochure is bordered in black, uses the same print and
layout as the Park Service, and is confusing a lot of people, who believe
that the park already exists.

All that's really needed to create it are two more basic steps: to have
Congress purchase the vast tracts of land currently owned by international
speculative timber and paper companies and then to have the U.S. President, any president, dedicate that land as a park.

The text of the brochure is eloquent, the photographs properly amazing: autumn leaves in the Maine woods, pitched tents, hunting and snowmobiling (within designated National Preserve areas), snowshoe trekking,
old-growth forests, whitewater canoeing, climbing, lodges . . . Throughout the text, the proposed boundaries and points of interest, and activities
are treated as fact that has already occurred. Only on the back side, in small
text, is it explained that "today Maine Woods National Park is only a vision.
But you can help make this vision a reality."

The heart of the Maine Woods deserves to be protected as a new national
park. As a national park, the natural wonders of the region would belong to

the public. The land, water, and wildlife would be preserved as priceless parts of our national heritage, safe from exploitation. Vast wildland would offer solitude and spiritual renewal to civilization-weary people. And the Park would strengthen local economies by creating jobs in tourism, recreation, and ecological restoration.

The vision of Maine Woods National Park can inspire people across America. With strong public support, this park can become a reality. Several of the existing large land owners are now marketing their "non-strategic" holdings within the proposed park boundaries. Most of the remaining lands are held by a handful of large land owners that are likely to be willing sellers in the future. Based on prevailing prices, these lands would cost approximately $100 to $300 per acre to acquire. For less than the cost of one B-1 stealth bomber, the American people could create a great Maine Woods National Park for this and future generations.

Like all generations past, our children and our grandchildren should be able to hear the cry of the loon, the song of the hermit thrush, and the howl of the wolf.

The howl of the wolf: part and parcel of the concept of wildness. Yet where is the wolf in this story? It has not yet crossed the border, or rather, has not yet survived the border crossing. In Montana, it was deer, and a softening of public attitudes, that seduced the wolf back. Here, likewise, it is nothing less than heart's desire. Stacks of paper, petitions, entreaties, summon the wolf. And I think that will be enough to bring the wolf filtering down through the forest. It's very strange.

RESTORE has prepared a ten-point "Question and Answer" sheet that addresses the most common concerns.

Would private or state lands be taken? No! Under this proposal, park land would be acquired by the public from willing sellers only. The focus would be on the large paper and timber company holdings, which have been changing ownership frequently in recent years. Owners of smaller parcels could keep or sell their property as they choose. Towns and year-round homes would not be part of the park. Private camps now under lease would continue under long-term leases or be acquired from willing sellers. Access to private inholdings would also continue.

Would crowds and over-development be a problem? Within the proposed park, wildland values would be protected. No new commercial development would be allowed on park land. Crowds would be minimized by spreading visitors over an area much larger than Acadia National Park or Baxter State Park. Outside the park, local towns could take advantage of new economic opportunities. As has been done at existing national parks, towns near the proposed park and park staff could work together to guide growth, prevent unwise development, and protect the local quality of life. The proposed feasibility study would assess these important issues.

What kind of uses would be allowed in the park? The proposed park would actually be a park and *preserve*, guaranteeing public access for traditional recreation. The entire area would be open to hiking, camping, canoeing, rafting, fishing, and most other recreational uses. Hunting, trapping, and snowmobiling would continue in the preserve portion, as they do in other national preserves. The size and location of park and preserve areas would be decided by the public during the study process.

And, as ever, the all-important question of our culture:

How would a national park affect the economy? A new national park would offer many economic benefits. The park would: (1) supplement the troubled forest products industry while leaving four-fifths of Maine's commercial timberland unaffected; (2) create jobs restoring damaged forests, rehabilitating wildlife habitat, and managing the park; (3) draw new businesses and permanent residents seeking a healthy natural environment; and (4) expand the tourism economy (Acadia National Park brings over $160 million per year to that region). It could also increase the tax base, since federal payments are typically higher than current property tax payments by corporate land owners. The proposed park feasibility study would address these critical issues.

It's a very slick and polished, practiced campaign, incorporating the spiritual, physical, emotional, economic, historic, and ecological needs of our present culture. I think the park, the land, will get protected, and that—along with New York's Adirondack Park, and the swamps and bug-thickets of Vermont's Northeast Kingdom—substantial blocks of wildness will be protected in the East.

I think that someday soon, maybe in the next generation, young men and women who rise into the post-adolescent need to disperse and seek and explore new wild country will not have to strike out West, as so many of them currently do, but will find the spirit of wilderness much closer to home.

What effect would that have on such young people? How would their hearts be different from the past generations who have felt the need to light out for the far territory, traveling north and west, running until they can't run?

Is it blasphemy against the West to suppose that such a staying-at-home, as opposed to distant flight, might craft a slightly different heart—perhaps even a heart fiercer and calmer than the current flightstrong, panicked, desperate population of wilderness advocates who have come from the East to fight for the last of the West?

And those young people being born into the West now and raised at the edge of the West's diminishing wilderness (even as the East's wilderness is expanding)—what lessons, models, stories, encouragements, can we put down for them now, as they wander out into the near future?

Pilgrims, all.

While utterly in awe of DeBoer's commitment, talent, endurance, and resolve, I had to confess to her and to Kellett that I have some questions about both projects—the park, and the intentional reintroduction of wolves, rather than letting them come back on their own through the filter of natural selection.

Like many environmentalists, I favor natural recolonization of endangered species wherever possible—fixing whatever localized problems led to the species' demise in the first place, rather than necessarily bringing in new recruits by helicopter.

I prefer an approach that looks at the protection of the critical habitat required by an endangered species, as well as all the other plant and animal relationships that species depends upon, rather than focusing disproportionately on the one isolated species; for in a wild and healthy landscape, there are no isolated species.

The "helicopter" approach, I fear, runs the risk of letting us inure ourselves to the cataclysmic pace of ongoing extinctions, lulling us into the belief that we can patch any leak in the ark; that we can buy or shop our way out of any jam; that we can avoid looking directly at the root of the problem, which is, again and again for so many of our threatened and endangered and sensitive species, the vanishing of wilderness and the fragmentation of important habitat. The halves being cut to quarters, the quarters into sixteenths. The subdivision of wilderness, and the proliferation of the homogenous.

I sometimes feel guilty for believing this as strongly as I do, knowing that in many ways my own perspective is one of privilege unequaled anywhere in the Lower Forty-Eight states, for my valley is the only one where nothing has yet gone extinct—not since the time of the last ice. (Nonetheless, many of my valley's populations are now at extinction's edge, hammered down to single- or double-digit populations by relentless industrial road-building and clearcutting: less than twenty grizzlies—we lost *five* in one year—five wolves, twelve pair of bull trout, one occasional woodland caribou, a handful of lynx and wolverine.)

Would I feel differently, I wonder, if any certain species in my lush, rank valley were to blink out—the Coeur d'Alene salamander, the redband trout, the white sturgeon, the moose, or the great gray owl, fisher, or wolverine? The grizzly?

If the grizzlies were to disappear—as many independent biologists are projecting can happen in the next century, beginning in the lowest-elevation habitats first, of which Yaak is the lowest—would I change my tune? Rather than requesting a hundred-year conservation plan aimed at steadily restoring the gone-away critical habitat, and then allowing the remnant species to slowly re-inhabit those recovered lands (like the drift of seed pods), with the most "correct," land-fitted individuals succeeding against the filter of natural selection—would I instead be clamoring as loudly as anyone else to bring the grizzlies back, to bring them back now?

I don't know. In any event, DeBoer was able in short order to convince me of the merits of wolf relocation, rather than waiting for natural recolonization.

As she explained it to me, an impenetrable and unnatural gauntlet lies between Canada's existing wolf population, and the United States' available wolf habitat. In this instance, getting wolves to come back into the United States is not so much a matter of waiting to recover country that is sufficiently wild to receive them—the landscape is ready. Rather, immigrant wolves would need to surmount that nearly impermeable plug that exists as a thirty-mile-wide band of sheep-farming country just across the border into Canada.

Almost any wolf that is lured into this zone is trapped or shot. (It's strange to think of a wall of sheep, thirty miles thick, being able to stop a wolf, but that's how it works.) In the last ninety years, only one wolf is known to have made it into Maine from Quebec; and that wolf, a 70-pound gray female, after having miraculously survived the Quebec gauntlet, ended up being shot by a "hunter" in Maine in 1994, when it wandered over to some bait the "hunter" had set out in hopes of luring in a black bear.

The St. Lawrence River and Seaway is thought by some to be an obstacle—it hardly ever freezes over, and in the places where it does, the development of cities and towns has been rampant. But this question, as well as the Quebec plug of sheep, might be irrelevant, when placed in a larger context. It's possible that the eastern timber wolves in Quebec are going extinct, and their risk of extinction in this part of Canada is another reason some should be relocated to their former habitat in the United States and given full protection.

Across the border in Ontario, the only true protection wolves have lies in the Algonquin Park. Even there, wolves are killed any time they step outside the park boundaries.

In the Parc Provincial des Laurentides, in Quebec, perhaps only thirty wolves reside in an 8,000-square-kilometer park, which is a very low population for that size and quality of habitat. The numbers are so low that even the Quebec trappers' association is lobbying the government for a ban on wolf trapping until numbers recover—but, in a stranger twist, the government is resisting the trappers' request, believing that that beaten-down wolf population is somehow limiting one of the province's cash crops, caribou. The government desires that the wolf population be knocked down even further. How low do they want to go? To seven wolves, or six wolves, or five? It's the twenty-first century. Have we truly become so weak and small and scared that the most room we have for wolves anywhere is a number that we can count on one hand?

So I can see clearly the logic of capturing and relocating wild Canadian wolves (not from Algonquin Park!) into unoccupied prime habitat over on the U.S. side. The ethics of such an operation are dubious at best, but we gave up any chance of being pure in this game quite a long time ago. If it helps the wolves, and helps the wilderness, I am probably going to be in favor of it.

Regarding wolf habitat in the Northeast, dozens of scientific studies and fancy color-coded GIS (global-imaging system) maps are springing up in scientific circles, maps and studies that give the impression to me at least of providing stepping stones for the coming reintroduction. To pretend to be able to predict fully where the wolves will go, however, or how they will behave, would be at best arrogant; and yet to ignore the best projections we can make would be irresponsible.

Hence, in preparation for the wolf's return, we have scientific papers with titles such as "An Assessment of Potential Habitat for Eastern Timber Wolves in the Northeastern United States and Connectivity with Occupied Habitat in Southeastern Canada," and thick, exhaustive surveys such as "Wolf Restoration in the Adirondacks? The Questions of Local Residents." These studies have been many years in the making, since the U.S. Fish and Wildlife Service first began seriously considering the question in 1992. One particularly impressive map, titled "Distribution of Occupied and Potential Habitat for Eastern Timber Wolves in Northeastern North America," is color-coded to show potential core habitat, potential dispersal habitat, and potential unsuitable habitat.

In the hypnotic way that maps have, these maps capture you, as a dream does: the way the abstract is rendered suddenly into certainty; the way new boundaries and borders are both loosened and established, guideposts for a situation that seemed previously ungraspable.

It's arrogant and yet magical, the way we can distill our perception of 26 million acres of seething forest and mountain and river and swamp onto one 8½-by-11-inch sheet of paper, with the peregrinations of hypothetical wolves-of-the-future color-coded into those unseen clefts and folds of topography.

For now, however, the only alternative to that single sheet of paper would be to ignore wolves completely, which we are certainly not willing to do, nor should we. Maps and dreams have always been inextricably linked; they are but part of the process, in which we can participate more fully in the wolves' return.

I support the idea, and ideal, of the Maine Woods National Park, but I also have concerns. Living shoulder to shoulder with hunter-gatherer types—some of whom I respect, others whom I don't, but nonetheless, committing my life to this one place, this one community—has sanded and rounded me into more of a moderate than I might ever have been able to imagine.

Having spent my whole life as an extractor—not "just" through my steady consumption as a human being, but as an oil and gas geologist, as a hunter of deer, elk, and grouse, as a producer of books and stories that use paper—I am acutely aware of the hugeness of my weight, each of our weights, in the world.

Such awareness, I think, does not counter but rather gives rise to the reason, the need, to protect our last wilderness. And so I am attentive to the idea of this big-ass Maine Woods National Park and Preserve, and understand on all levels—rational, scientific, emotional, intuitive—the logic of it.

But I, and others, have been waging such a small and isolated campaign for so long—trying to gain some kind of permanent protection for only 175,000 acres, scattered in thirteen little pockets across one hard-hit little valley—that at first the idea of creating, or rather, protecting, a 3.2-million-acre park is beyond my grasp.

When it seems like you're wrestling with the impossible, it's quite a task, mentally, to consider multiplying that challenge by a factor of almost twenty.

As well, I can't help but be struck by the change such an act might bring to the handful of ghost-hamlets cast here and there through the woods within a stone's throw of the proposed park. No private land would be taken for the Maine Woods Park—the ownership would be comprised only of lands purchased from willing sellers, most notably the gargantuan paper companies who own blocks of land measuring in the hundreds of thousands of acres, and whose designs upon the land are to devour it—but still, as a shy hermit, a recluse of the dark wet woods, I can't help but think of how much I hate the sight of those damn Winnebagos. I wouldn't want that for the Yaak. I might even continue to take clearcuts, before that.

Maybe I'm on the wrong track here, but it seems to me that a massive, contagious, chronic disrespect often arises, in which the local residents of a tourist destination, in being gawked at—in being *seen*—are stripped of the private dignity each of us holds—and that furthermore, the touring pilgrims come seeking to take: in some ways more ravenous than the more classic despoilers.

It's a subtle thing, perhaps, but subtle things can be strong. The classical extractor comes to the land focused, knowing more or less what he or she is searching for—a ton of ore, a load of timber, a meadow of hay—and the exchange is made, the labor for the product; sometimes in sustainable, respectful fashion, though other times not.

For the pilgrim, however, there is perhaps something dangerous in what has become too often the modern approach to landscape. Under the seduction of advertising, the pilgrim can be tempted to come to the landscape every bit as ravenous as any earth-mover, due in large part to hyperexpectations that are as grand and vast as they are abstract and unknowable; expectations that speak to some empty, hungry place within each pilgrim.

The pilgrim approaches a landscape expecting his or her due. Believing that a due exists. The money has already been spent; the pilgrim has arrived.

Because the name of the thing is not known, the pilgrim scours everything with his or her view, looks far and wide, almost frantic, unseeing; rarely does he or she focus.

Is such an approach harmful to community or landscape? In a way that I cannot figure out, I sense that it is.

I'm not saying I think the Maine Woods National Park is a bad idea. What I'm trying to approach is the notion that (perhaps most gluttonously of all), I want it all: vast wild cores of country in which neither industrial recreation or extraction is practiced, places where the animals of the forest are still secure in the world, still filled with the uncompromised grace that has always been in their possession; a world in which, if a man or woman or child is to enter, it is with the respect and dignity, the humility, of the traveler on foot—the true pilgrim.

That some of us are aged or infirm and can only make such journeys in our minds does not justify erasing these last wild cores by putting roads through their hearts, and thus by making them into places that are just like all the other places.

And from those anchors of wilderness, then, I want more: I want the rural communities of people to survive, hovering with their tenuous relationships to the land that are so different from those of the rest of the world's. The dangers of a national park in northern Maine are clear; chief among them that the handful of hamlets scattered throughout the woods might become overrun with garish neon glitz of advertiso-rama.

Still, I can envision a plan that would guard those scratch-living rural communities as carefully as we would an active wolf den, a heron rookery, a nesting bald eagle. I want to dream and believe in a story in which sustainable forestry could still be practiced in places in a manner that would not harm the forest, and which might, in a mosaic of patches and places, help the deer by setting certain small groves back into earlier successional stages: not clearcuts, but openings of light, clearings, in which the deer could then feed upon the new emergent forbs, whereupon the wolves would then feed upon the deer, so that in effect the wolves were feeding upon those slashes of light.

It's five below zero when I leave Concord, New Hampshire, snow scrunching underfoot and stars flashing, and drive north and west into Vermont, driving through the night toward Sue Morse's tree farm. She is a professional tracker and a forester, and will be leading a class the next day on a walk through her wild woods, just south of the Northeast Kingdom. Morse lives in country that wolves will probably one day reinhabit, not simply because human beings desire it, but because it was once partly their country before. The shape of the land is still the same, has not changed much at all since the last glaciers; and once again, like a tide, wolves will slide back down over the hills and mountains and through the forests, laying down tracks upon their own ghosts.

Driving through the hills, up knolls and swales, through the snowy darkness, along the winding backroads past sleeping hamlets, it's hard to imagine even a single wolf navigating its way secretly through this countryside. But perhaps it could. Perhaps if it kept traveling, it could move through these sleeping woods with care.

Sue Morse radiates something, I'm not sure what. You could call it health, strength, vigor, even wildness. Whatever it is, it's vital. Her short strawberry-blonde hair is burnished as if from much sun, and in the thirty-below windchill this crystal-bright March morning, her face is ruddy, which gives her broad smile extra cheer. It's clear that this morning's joy comes from the prospect of getting to go out-of-doors today, weather be damned, to simply be in the woods. It is always a great and specific joy to be in the presence of one who finds such good cheer in the midst of— *because of*— harsh or inclement weather.

Sue takes people on woods-walks, and teaches them tracking, several times a year. She's helping coordinate a training program in Vermont, called Keeping Track, wherein volunteers conduct wildlife inventories in the woods around their homes. Across the years, these inventories will provide the state with a bank of reliable biological data and trends that could not otherwise have been obtained.

Sue passes out walking sticks to the couple dozen participants, and we start up the mountain, shuffling through the knee-deep snow. The forest is covered with snow from the day before, glaringly brilliant in the morning sun, and the sky possesses that depth of blue that you see only against new snow, and only at temperatures well below zero.

Because she brings so many people on these walks, over the course of a year, Sue asks that we take extra care to stay in single file and, where possible, to stay in each other's footprints, to keep from bruising or trampling dormant plants and seedlings beneath the snow.

This isn't to say that humans haven't been active in this forest. Sue owns a couple hundred acres of these woods, and she's logged them selectively, with care and pride, always toward improving, as she can, the health and vigor of the aging forest. She has a hunter's eye for which trees to leave, which to take, and which ones to let grow a bit older, just as a hunter might pass on a certain deer, even a nice deer, as long as that deer was still in its prime and growing vigorously. (Sue is an avid bowhunter.)

She also recognizes the full value of a dead or dying tree, an aging giant, to be left in the woods forever. She knows her forest intimately—the comings and goings of the seasons, the curves of the land, and the qualities of the soil in each little fold and pocket.

It gives me solace to be in her company, to be walking along behind the aura of her strength and confidence, her passionate moderation. So often, when I am seated at the computer, pounding away at the keyboard and feeling under the activist's siege, I feel that I am becoming ever-more a lightweight, always bending over backwards to take into account the opposition's position, considering their opinions and perspectives. I'll find those concerns factoring into my every thought, and because it feels like battle, and because I do not want battle, I'll often feel as if I'm being too radical —even when I'm writing what sounds to me like the sanest, most reasonable things. I feel like an extremist.

But then, whenever I get out into the woods and get up on some windy ridge, and look out not at some computer screen but at the rise and fall of the landscape I know so intimately—such a small piece of the West, but so important—always, always, I'll feel a thing like anguish: not that I have been too radical, but that I have been too moderate, in the face of what is at stake, and the dangers surrounding and encroaching on these last wildlands.

Listening to Sue talk about her hunting experiences, and her belief that we can still have "wilderness *and* logging, rather than wilderness *or* logging," inspires me. Her confidence, as well as the authority of her experience, inspires me to believe that the middle path *can* be had, just barely, if we hurry, and that we can still save the last corners of untouched, unprotected wilderness, and repair the injured lands in between.

It's a gamble: cutting trees to save them; cutting trees to save forests; cutting trees to try to save threatened and endangered species, rather than simply committing to a five-hundred-year plan of "hands-off" and stepping back to let things sort themselves out with nature's customarily intricate and unfathomable grace. But it's a gamble I believe in. It's a pleasure to find the company of another who feels the same way.

It's astounding, really, to meet someone so similar to one's self, in so many ways, and yet who seems upright in the world—so unfloundering.

Is the East only a less frantic reflection of the West, or the West a more panicked reflection of the East, as more and more of the West's wild treasure is spoiled, ruined?

Aren't all hearts pretty much the same?

For my part, I can only laugh. The face in the mirror over here isn't the same, but nonetheless, other mirrors seem to be reflecting the same hungers and desires.

Sue Morse loves to build stone walls! She can be fierce and ornery with "the opposition," though infinitely patient with the students she is teaching. She loves books. She loves feathers, stones, driftwood, hides, bones: the physical beauties that anchor this world of spirits, the structures and frames for the sparks within.

How different, how individual, really, can any of us be from one another —especially when we love the same things?

The forests and fields, beaches and deserts of the world are far more individual than our various hearts can ever be. When we're lucky, we are but echoes and shadows of those landscapes. When we have destroyed them, our own lights will cease to exist; we will vanish, cut off from the thing that had for all the centuries and millennia before mirrored and shaped us, and shaped our hearts.

We haven't traveled far before we hit the first tracks; another living thing has passed before us. A young coyote has been scavenging a deer carcass, and Sue shows us its tracks. Loose deer hair is scattered here and there, and new-frozen scats punctuate the coyote's wandering sentence of tracks, scat wrapped also in deer hair and studded with pearls of jaw-cracked bones.

In an open area, Sue shows us other sentences that pass over and across the coyote's sentences: the featherswept strokes where a raven had swooped in for a bit of bone or flesh, or perhaps to visit the coyote.

I don't want to dwell overmuch on the notion of spirits—such thoughts and discussions almost always wither when inked onto paper, and are instead best explored, I think, in private cauldrons and eddies of each person's heart—but I can see how for Sue Morse, having the knowledge, experience, and ability she does, it might be a little like being able to look into the spirit worlds in motion all around us.

As she moves through the woods, noticing the residue of tiny and not-so-tiny passages of others, it must be for her like a kind of second vision. She is able to see the hidden past as clearly as you or I might were we to take a few steps into the future and then look back to view the present from some slight distance ahead.

Not much farther beyond the coyote tracks, we find fisher tracks. In the soft, deep snow, the prints are not very distinguishable. But the distinct, loping, sinuous carriage of the animal—only the wolverine is larger among the mustelids—identifies it. Sue makes a few hops of her own, imitating for us that weasel-like swimming motion with which the mustelids attack the world, as if part fish, part snake, part mammal.

Deeper still, we find bobcat tracks. The sunlight in the old interior forest of leafless beech and maple is dense, seeming somehow from another time; columns of buttery gold slant against the winter-pale bark of the trees, mixing with the deep shadows to create a yellow and black mosaic that is nearly identical to the coloration of bobcats; and again, it is like a kind of vision, seeing those tracks pass through that light and knowing that we are only hours or even minutes behind the real animal.

Sue even knows this bobcat by name; she knows its patterns, its reproductive history, its size and temperament. She has been hiking these two thousand acres—her grandfather's farm—almost daily for the last twenty-three years.

Loosely formed as they are, the tracks are hard to distinguish from those of a coyote. Sue has us follow the tracks backwards—she doesn't like to push an animal ahead of her, is impeccably polite and courteous that way—and finally we come to a spot of diagnosis, a spray of fresh urine, which she instructs us to kneel down and smell. *Cat.*

As quoted in a recent *Audubon* article by David Dobbs, Morse explains the strategy behind her Keeping Track classes: there is scientific as well as activist value in having citizens gathering data from their own home terrain throughout the state, recording the patterns and occurrences, the comings and goings of things, with a degree of coverage that no state or federal budget could otherwise ever have afforded.

"Only by doing this sort of homework—literally keeping track of wildlife use—can we know a given habitat's true value," she has said. "And only by knowing its value can we work effectively to conserve it."

The place where we're walking is, she says, a case in point. "Twenty

years ago, if we had discussed as a town what to do with these lands, [all] I would have said was, 'Whoa! We can't build a Wal-Mart here! These woods are too pretty! I've seen moose here! This is special!'"

Now, however, with the weight of her documentation behind her, she can say, "The area functions as core habitat for bear, bobcat, fisher, mink, moose, and otter, and it's essential to the health of those species and everything they draw on. It's a big difference."

It's brutally cold back in the shade of the forest. Everyone is flapping their gloved hands and arms like stunned chickens, laboring unsuccessfully to stay just this side of suffering—it is only the beauty of the old forest that allows us to push back the pain in our toes and fingertips even remotely. Amazingly, Sue Morse is striding along bare-handed. She's breaking trail for us, so I'm sure her blood is running a good bit harder and hotter, and you can tell just looking at her that she's thrilled, always thrilled, to be in the woods. But the windchill is still way below zero, and I can't help but marvel at all the different tones and variations of wildness in the world, including among our own kind.

We come out of the old bottom and ascend a ridge, and begin to find more bear sign. The great beeches and birch trees have been climbed and clawed, the smooth bark latticed with the ancient hieroglyphics of bears past and present. Sue points out the ladder-stacked clusters of branches in some of the higher trees: large nests that the bears have built in which to lounge, as you or I would lie in a hammock, as they strip and browse upon the smaller, greener limbs of the trees—eating those young limbs the way you or I might pluck grapes from their vines.

We hike all day, crossing the occasional tracks of bobcat, otter, fisher, deer; ascending cliffs and descending along rushing creeks, until it seems certain that we have left Vermont and are in some other, wilder place.

Nearing the end of our loop, returning to Sue's cabin, we pass along the boundary lines of her grandfather's property, which is punctuated by immense, ancient maples—"witness" trees that were not cut but blazed to serve as survey markers—trees that were substantial even back in the days of the land grants, two hundred years ago.

We pass the rusting hulk of a crumpled wood stove, moss-anchored, being swallowed by the forest. At the beginning of the century, Sue tell us, almost all of the forest we're walking through right now was cutover fields and denuded sheep pastures. There were accounts of wolves being so hungry back then that they would come out into the meadows and gnaw on the maize from farmers' fields. Then—like the forest itself—the wolves went away for a while.

Now the woods are coming back, and the wolves will be coming back, and barely a single human lifetime has passed, two humans' lifetimes.

Our little cycles are nothing, compared to the world's. Our greatest dramas are like those of the mayfly. We spend a few weeks scrounging gravel and building a little edifice. Then some force summons us—we ascend to warmth and light, spin with violent beauty and a strange mix of joy, confusion, instinct, and desire—then fall from the sky, back to the river, or to the gravel bar at river's edge.

Farther downstream, all around us, other mayflies rise, like the lifting and falling of the world's thread-like pistons. As if, daily, all disappears; is buried; is lifted up; returns.

That night we dine in Sue's cabin, which is filled with more books than I've ever seen in a cabin. The stuff of her life surrounds her—spotting scopes, telescopes, binoculars, hides, pelts, antlers, rocks, feathers, and bones—all beautiful, testimony to the great grace and design of the world. But what strikes me most is the books: shelves of books, stacks of books, books on tables, and books resting in chairs, half-read. What magnificent paradoxes we are, for a woman as fully engaged in the out-of-doors as Sue Morse is, to also have her life, her spirit, so immersed in ideas, the invisible and abstract concepts of words and sentences, each of which always drifts a slightly different path for each individual reader, rather than possessing the clarity of a track in snow, or a claw mark upon a tree.

We dine on hot lamb chops and cold beer, sitting cross-legged around the woodstove, surrounded by those great books of poetry and natural history, novels and field notebooks, journals and scientific books; and we discuss, as it seems environmentalists are always doing, the future.

Sue raises organic, free-range sheep on her farm. We're eating one of her lambs tonight. She doesn't trap coyotes, but instead uses guard dogs to protect the sheep. She wants the wolves to come back. It seems at first a wild and wonderful paradox, her wanting both sheep and wolves, but then, after a day in the richness of her country, it comes to seem natural. There's still—with resolve—barely time and space for both.

It's her job to take care of the sheep; it's her job to take care of the wolves. *Stewardship.*

The scholars, professors, and activists Chris Klyza and John Elder have joined us, and Chris, an intense and intelligent young man, talks literature for a while, until, like magma seeping from vents in the earth, numbers and statistics instead begin to issue from him. The fact that 20 percent of our population is already experiencing a shortage in fresh water supplies, and that in the next twenty-five years, over 30 percent will be lacking or stressed.

The fact that the world's human population is expected to double in the next fifty years.

The fact that sixty million people live within a day's drive of where we're sitting.

Are we—moderate environmentalists—blind to the reality of the situation? Are we blinding ourselves, perhaps as a defense against the awful weight of the truth, while we move through the world neck deep in our own ferocious consumption, and continue to proclaim that hope for the future still lies in phrases like *sustainability, stewardship,* and *moderation*— that indeed, we can (and must) have more wilderness, and at the same time more (but smarter) logging?

Can we still find a middle way to carry forward these two desires, these two needs, rather than abandoning one or the other?

Probably, for one or two more years. This is probably the last cusp of time in which an environmentalist can be afforded the indulgence of believing such things. Surely it is our last last chance.

John Elder agrees. "The game is now," he says. He's talking about Vermont, and the rest of New England. "The pressures to develop are going to become even more intense, in just the next few years," he says. The booming economy, the astounding affluence, the eagerness of the large landholding timber companies to divide their land into the smallest possible fragments . . .

Where are the wolves in this story? They're still a couple hundred miles north; still out in the future. An unconnected triangle still separates the wolves, this landscape, and the desire in our hearts for the wolves to move back onto this landscape.

But it seems that the pieces are beginning to move closer together; almost like a thing that is getting up and beginning to walk.

And later that evening, as we're talking about innovations in forestry practices, value-added products, sustainability—the night's frigid silence is broken by a coyote's howl.

It's so close and so loud that we can hear it even through the walls of the cabin, through the old logs from the forest, and through the closed windows.

"It sounds like a *wolf*," I say, and I look at Sue with some suspicion— wondering if she's not keeping some sort of secret.

"The coyotes are really big around here," she says. "They average about seventy pounds." (The size of a young or small wolf.)

The four of us wander out into the depth of the night. It's like walking out into a dark ocean. John and Chris are marveling at the stars. We truly could be at the edge of a wilderness. Places like this, pockets and patches, still exist in the East.

The coyote's howls are wild and fierce and strong: *so close.*

One of us murmurs something, and the coyote hears it, and falls silent. We wait for it to resume, but it's too smart, too wily. For us, it's too cold outside; we linger as long as we can, a few moments longer, stamping our feet and waiting and listening.

Sparks and the smoke of burning trees rise from Sue's chimney, up into the leafless canopy of the trees; higher still, then, the sparks seem to be drifting through the stars themselves.

We stand a few seconds longer in the icy night, just outside the glowing yellow cabin, caught perfectly between domesticity and wildness, as if that's where we're most meant to belong.

If so, when there is no more wildness, or no more wilderness, what shall we do? What will become of us?

It's so cold out that we can see the long crystals of ice slashing and glimmering in the starlight as they mix in a slight stirring breeze. They look like teeth, long sharp teeth, falling upon the land, gnawing and tearing at the land, but causing no damage.

So people's hearts are changing, opening up, it seems, to make a nest for wolves—not just Sue Morse's and Kristin DeBoer's hearts, but millions of people's. The land is opening back up, too, even if briefly. Sometimes the story begins in Vermont, and sometimes it begins in Montana; other times it washes back and forth in waves.

I think that anyone who tells you that he or she knows what the Northeast wolves will do upon their reintroduction is kidding themself. They will do what they have done everywhere else, which is to combine the predictable with the unpredictable. They will feast on beaver, raccoons, deer and moose and sheep, and perhaps on dogs and foxes, as well as coyotes.

They might or might not try to breed with the coyotes, which could lead to the washing out of one thing, "true" wolves, and the braiding or creation of another thing, "new" wolves.

Some of the wolves will prosper, others will starve. Some will probably filter over from Canada; others will be shot, or killed by traffic. Some will slip their radio collars and never be caught again.

Some will be seen by tourists. The tourists will come to hear them sing, and to see the wolves' big prints in the snow and the mud; to examine, in the warm light of summer, the bright cracked and scattered bones of deer and moose that the wolves have fed upon, lying here and there in the forest.

The tourists and pilgrims, as well as the residents of the Northeast, might look up at a shining black raven as it flies overhead and understand that some of the deer meat went into that raven as it scavenged the kill, and begin to consider the new braids being woven—the wild eating the wild, living on the wild.

As Champion and Plum Creek ate Montana, so too are they eating Vermont and Maine. In northern Maine—in the heart of wolf country—more of the same old boring, unoriginal game of Montana cut-and-run is going on.

Plum Creek has just purchased a big wad of forest land from another timber company, and is forming a new spin-off group, Plum Creek Realty. Conservationists are rightly fearful, believing that history will repeat itself, that instead of simply cutting and running, as large timber companies have done in the past, Plum Creek will cut, run, *then* sell—and sell, and sell, and sell.

As if putting salt in the wound, when the purchase in Maine was announced, Montana governor Marc Racicot sent a letter to the governor of Maine, Angus King, informing him what a good corporate neighbor Plum Creek had been, and what responsible stewards of the land they were.

This prompted a letter by the Montana Wilderness Association that expressed dismay and surprise at "the uncritical and unbalanced portrait of Plum Creek" that the Montana governor sketched.

> To those who have followed Plum Creek's management of natural resources and lands in Montana, your glowing comments can only be understood as a reflection of a personal regard . . . for Plum Creek's officers and directors, not as an official appraisal of and acquiescence to Plum Creek's actions in Montana.
>
> If we had written the letter to Governor King, we would have pointed out that many reasonable people in Montana hold a very different view of Plum Creek's activities. In particular, we would point out that an honest assessment of Plum Creek's activities begins with recognizing that it has employed some very personable individuals in recent years who have honed a highly sophisticated public relations effort on a scale previously unknown in Montana.
>
> In 1989, Plum Creek spokesman Bill Parsons spoke bluntly and truthfully when he said, "We have never said we were on a sustained-yield program, and we have never been on a sustained-yield program."
>
> We must also report that the company's . . . forestry also does not measure up to the claim that the company is a national leader in forestry management. On a minority portion of their current logging jobs, Plum Creek does a good job, leaving species and structural diversity through careful logging. These lands largely are in highly visible areas.
>
> In most places since Plum Creek's "environmental forestry" was an-

nounced, however, the company's selective logging amounts to little more than high-grade logging. Almost universally outside Plum Creek's immediate circle (and sometimes within it), government and private foresters, loggers and biologists speak derisively about Plum Creek's environmental forestry. Some . . . say it's "take the best and leave the rest."

We regret that your letter to Gov. King presented an uncritical and unbalanced report on Plum Creek's activities. Not every person in Montana, and not even every logger in Montana, shares your sense of enthusiasm about Plum Creek's activities here or expansion elsewhere. As a state, we . . . continue to pay the price for Plum Creek's activities.

A nonprofit organization called The Conservation Fund (much like the Montana Land Reliance) is buying 133,000 acres of cutover Champion land in the state of Vermont that could otherwise be fed to the hounds of development; of that total, they intend to sell 26,000 acres in Vermont's Northeast Kingdom to the U.S. Fish and Wildlife Service, who will then be able to establish a federal refuge. (I wish such dollars were available in northwest Montana, to buy the Plum Creek lands for sale in the Fisher River country. But I understand that just as the West has more wilderness, the East has more money, and there are only so many conservation dollars available for the whole country.)

These "little" Vermont tracts of 26,000 acres and 133,000 acres pale beside South Africa Pulp & Paper Industries Ltd.'s plans to sell 911,000 acres of forest in northern Maine. (About half of that land lies within the proposed boundaries of the Maine Woods National Park.)

Additionally, it's reported that Bowater Inc. is planning to sell or close its century-old paper mill in Millinocket, which would leave vulnerable— or available—another 2 million acres owned by Bowater.

Where are the wolves? Where do they find the seams between the likes of South African Pulp & Paper Industries and Plum Creek? Between the Republican-led 104th Congress' "Contract On America" and thirty-five years of failed wilderness legislation in the Yaak—between hope and despair—where, if at all, do the wolves of the future find a home?

Is it really true that we have two competing gods in this country: the god of money and the god of spirit, which for many is typified by or most easily experienced when one is out upon a wild landscape?

I think increasing numbers of us are becoming almost frantic at the war these two gods are waging within our hearts.

I think the government should buy that land up in Maine that's for sale—should buy it for the people, and for the future. I think we should buy the timber companies' land back, where we can, in Vermont, too, and in Montana—the old railroad lands deeded by Congress that eventually ended up in the hands of Champion and Plum Creek.

Some folks propose simply wresting these lands back from the corporations, via the same kind of act of Congress that wrested them from the public and gave them to the corporations in the first place. I don't know about any of that. I just know we need, intensely, more wild places, in both Vermont and Montana, and that if we don't set up some system for the preservation and protection of such places, the world itself will most assuredly not do it for us.

If we are going to continue to have the monolithic, supremely powerful corporations in our lives, then it seems we should have more wilderness, too; places where people can go, on our hands and knees, if necessary; places where the corporations cannot follow.

Perhaps I am talking like a crazy man. But some days even the idea of such places gives me enough strength to bear the abuses of a society infested by corporate values, a culture of so-called style over substance, in which a single glance at the surface of a thing assigns that thing its cultural value, rather than the judgment being based upon any examination of what is within. Some days, when even the *idea* of wilderness is enough to give me a moment's peace in the midst of this society of rampant quantification, in which everything is deeply commodified—a world in which at nearly every turn some bright and colorful voice or picture is urging us, begging us, to buy something: promising that with that one next purchase all the horrors of emptiness and loneliness will be staunched and assuaged.

The very idea of wilderness, much less the reality of its presence, gives me relief in a world in which our spirits are oppressed not only by the omnipresence of monetary values, but by the fact that any discussion that veers away from the dollar and toward the unnameable sacredness of what were once the most ordinary things—rivers, wild forests, wild animals, or the uncontrollable, passionate swelling of one's own emotions—is treated with sarcasm or irony at best, ridicule at worst.

Again, for me, wilderness is one of the greatest antidotes to such a world.

And if we are to still have corporations, and if we are somehow, miraculously, to have wilderness, can we not also still perhaps have a middle way, even at this late date? I wonder if it is still possible to achieve the place and condition about which Barry Lopez has written, in which "One of the great dreams of man must be to find some place between the extreme of nature and civilization where it is possible to live without regret."

Again and again, watching a raven fly through the bright summer sky, I'll have that repeated thought; that wolves are comprised of light. The plants that deer eat grow in that sunlight, convert by the mysteries of life's code that sunlight into cell material, which the deer eat, before in turn being eaten by the wolves. (After the wolves catch the deer, the ravens too share, briefly, in this feast of light.)

In this manner, some dreamy days, it will seem to me that the Sun, and its slashes of bold, bright light, holds the wolves aloft; even knowing, as I do, that the wolves prefer also the darker, farther woods.

One of the potential ironies of both Vermont and Montana, with regard to the management of forests, is that is is possibile that some limited timber harvest in some places might actually benefit wolves (if people agreed not to shoot the wolves, or if the landscape were secure enough from people to make the wolves' safety guaranteed). Silvicultural prescriptions such as fire, or selective thinning, can benefit one of the wolves' prey species; white-tailed deer.

The creation, or re-creation, of patterns of small mosaics of openings in the forest, such as those that existed historically, could provide more young nutritious forage for deer, which might then allow a higher population, which might then allow the wolves more available food supply, in the form of deer. (It's important to remember that not every species or population is limited by food; in the Yaak, for instance, elk need greater security— hiding areas—more than they need additional food.) This notion of small openings in the forest to possibly help deer obviously has the dangerous potential for abuse, corruption, misapplication.

In Montana, creation of more forage for big game has become one of the dominant excuses for the vast clearcuts that have turned some forests upside down, converting them from forests of old growth into plantations of young trees. It's almost the opposite part of the cycle, in the Yaak, for instance—the forest being carved ever-younger—while in Vermont and the Northeast, the forests are growing, finally, once again, steadily older.

I can't overemphasize the delicacy, the danger, of my even speaking this heart's consideration: that some logging can exist in wolf country, beyond the margins of wilderness.

Part of me understands that wild nature, even with her occasional seemingly volatile mood swings, is far more balanced than any system or design that can be replicated by the hand of humans; that a tree falling in the forest to a gust of wind, or disease, or an ice-storm, also creates beautiful mosaics of darkness and light that might serve wolves just fine, or better than fine. I am not saying that to have wolves we need logging, or even more deer. (Too many deer could tip the balance overboard, over-browsing and altering the forest even more drastically, and leading to disease and huge die-offs in the deer herd, so that in the end, as is too often the case with our management applications, we achieve the exact opposite of our desired effect.)

And certainly I am not speaking of logging in any form or fashion in the wilderness areas. When I accommodate the notion of a middle way, it is for a sustainable forestry reserve located between the wilderness and town. The wilderness should stay wild. All I mean to say is that sometimes when I look at a slash of light in the dark forest, or at a bold raven flying strong through the blue sky, it seems that I am almost looking at a wolf—or the future coming of a wolf.

Montana has the wolves, and yet there is very much the feel that we are living in the past—as if our wolves and other wildlands are but an echo or dwindling shadow of what was here before—whereas Vermont, and the rest of the North Woods, does not presently have the wolves, but has the feel of taking a brave step into the wild future.

Does everything travel in circles?

With a strange mixture of pleasure and envy, I examine publications by such organizations as the Vermont Natural Resource Council, and *Northern Woodlands*, as well as Vermont Family Forests (VFF), and literally dozens of other grassroots community projects cropping up in every watershed, every community, whose intent is to sew the wild woods back together.

Vermont is "blanketed" by forests—roughly 80 percent of the state's six million acres—and, completely unlike Montana, it is mostly (about 80 percent) privately owned. Vermonters refer to such woodlands as "family forests," and the bulk of these new grassroots groups are involved with securing conservation easements to protect the integrity of these lands and empowering the land owners to manage their lands in a sustainable fashion that is productive for wildlife and the integrity of the forest.

The 1997 Vermont Environmental Report quotes Wendell Berry on this subject: ". . . The two great ruiners of privately owned lands are ignorance and economic constraint." Almost anywhere in Vermont these days, it's possible to find "conservationists" and "loggers" who are speaking about the Wildlands Project's desire to create a system of ecological preserves, keeping that 26-million-acre ecosystem healthy and intact, while discussing in the same breath sustainable small-scale forestry practices.

Groups such as Vermont Family Forests (founded by Addison County forester David Brynn, of Middlebury) are already knee-deep in the second phase of a "smart wood" or "green" certification project in their area, such as the one envisioned by Steve Thompson. Since 1995, VFF has sponsored numerous workshops designed to provide "non-professionals" with tools to protect sustainable forestry. They have adopted a set of voluntary timber management practices designed to protect site productivity, water quality, and biological diversity. Their workshops have covered a wide variety of forest-related topics, such as portable sawmills, solar wood drying kilns, wildlife habitat, amphibians and reptiles, chain saw safety, riparian zone restoration, wood identification, timber grading, and water quality protection.

Thirty-one parcels of land have already participated in the "green certification" process. (In *Northern Woodlands* magazine, ads for rocking chairs, cabinets, bookcases, all brag with their FSC labels—Forest Stewardship Council—displayed prominently.) Other FSC goals regarding smart wood certification are:

1. To promote the use of sustainable forestry practices on family forests;
2. To improve the financial returns of family forests;
3. To increase the availability of sustainably produced, locally grown forest products to local wood product manufacturers; and
4. To develop an affordable model for independent "green" certification.

Actually, I'm envious as hell. If Addison County, Vermont, is able to pull together thirty-one different landowners across 170,000 acres for such a common goal, then surely we out West can pull together one common land owner, the U.S. Forest Service—which is to say, we-the-public—in similar fashion.

Referring to such imaginings in a recent letter to me, the district ranger up here called this kind of hope that of a "dreamer"—implying that he didn't believe it would happen, or was possible, or, I think, even worthy.

But it is happening in Vermont.

It's happening with such vigor and imagination that Vermont is bucking the national trend in forest products employment. By June 1997, the number of forest jobs in Vermont had increased, rather than decreased, for the first time in recent memory. According to the Vermont Forest Resources Advisory Council (a volunteer council that was, regrettably, disbanded by the state legislature), "the value of wages in forest products had eclipsed *all* other Vermont manufacturing sectors."

Helping drive this revitalization was—no surprise to classical economists—the efficiency of conservation. By focusing more on value-added manufacturing and marketing, the report discovered that "several mills are realizing productivity gains without increasing their demand for logs."

How far behind the curve of history are we in Montana? Despite being able to boast of our hundred wolves—a piddly number, really—we're still just decking together all our logs from a typical big timber sale on a national forest. A typical operation in Vermont, however, will sort its logs into several different piles, designed to get the top dollar for each species. A logging operation in Vermont might separate its logs into the following categories: white pine sawlogs, white pine pulp, spruce-fir sawlogs, spruce-fir pulp. White ash sawlogs might go to True Temper in Wallingford; sugar maple sawlogs to A. Johnson Company in Bristol; pallet logs of different hardwood species to Rutland Plywood; white birch sawlogs to George Adams Co. in Moscow; and veneer-quality logs in maple and yellow birch to Columbia Forest Products in Newport, Vermont.

In the meantime, in Montana, and in the Yaak, we're still just wadding them all onto the back of a single truck, or many trucks, and driving them off to Idaho or some other faraway place. Sending them, as ever, downstream, where someone else will add to their value, sometimes ten- or twenty-fold.

Maybe we're not smart enough for that yet. Or maybe we're just too damn lazy. Or maybe we just don't care yet about who or what comes after us.

These concerns seem a long way from the world when you are camped by a crackling fire in the north woods, or any deep woods, waiting for the sound of a wolf. What does it matter, really, in the greater scheme of things, whether a twelve-foot length of maple sells for $70 rather than $136? When your goal is to know one of the nearly impossible dreams of

this day and age—that time and place of quietness, of peace, and of biological fullness, of which the wolf, like the wolverine, caribou, and grizzly,
is one of the dominant and most dramatic components—what does all this
other ephemera matter?

How can rural politics, or for that matter, even Beltway policy making
and political favors, debts and paybacks, stack up against the depth of feeling far within you when you gaze into that fire or look up at the stars, the
same stars men and women have always looked at?

What I think and believe is that conversations like these are the ones
that prepare the way for the wolves' return. That these discussions, and the
policy that will surround them, are becoming a part of the landscape; and
that like any part of the landscape, they should not be ignored. Given the
immense and at times fluffy or romantic marketability to starved urban
dwellers for all things wolf, I have no doubt that a reintroduction process
could proceed in slam-dunk fashion, with policy driven by those incredibly
powerful cores of Eastern urban constituents. Vote *yes* or *no* for wolves.
End of story.

Simply going out and acquiring the land, and acquiring the wolves, and
then emptying the wolves out onto that land would be gratifying for me, as
would be a good meal—but it wouldn't be nurturing.

It wouldn't, to continue the metaphor, teach me how to grow my own
food and prepare my own meals for the rest of my life. To purchase the
land, nab the wolves, plunk them down, and then walk away, seems to me
too much like mere shopping; no more spiritual or noble then scratching
an itch or fulfilling an urge.

To accommodate, however, rather than vanquish, the rural communities hanging on at the edge of the wilderness—rural communities that,
like it or not, will be consumed anyway by the swell of the coming century,
with or without wolves—well, the challenge of that middle way intrigues
me, and I suspect that it appeals also to the glutton in me, wanting it all,
demanding it all. To me, it just feels right.

As well, the pragmatic considerations might convince you to turn for a
moment, only a moment, away from your gazing at those stars. We can
protect all the remaining eligible wild country as parks, monuments,
reserves, wilderness. But beyond that, know that the unprotected areas
will have to produce *something*; it's simply our nature, our destiny, to
tinker with them. To tinker with anything we can get our hands on. And
in these places—these gardens, these tinkering sheds, these stretches of

undeveloped country that are not quite yet either wild or tame—I'd still choose the making of bookshelves, the crafting of rocking chairs and cabinets, and sustainable logging, or sustainable agriculture and its attendant land ethics, its implicit demands that something be given *back* to the land, not simply taken, over the alternative, which is more parking lots, more traffic, more golf courses and hot-dog vendors, operations in which the dominant ethos is that of taking, not giving or sharing. Call me old-fashioned, but I know which I'd choose. And I know that those kinds of cultures, and those communities, are not going to survive without our help and support, even as many of their inhabitants spit at us, knowing only that we are environmentalists, and that we must be hated because they have been told to hate us.

How envious I am of Vermont, and all the tight-knit little community groups; all the pulling-together; all the commitment to problem-solving; all the *progressiveness*.

Back here, at home, I keep chasing this idea of middle way, even as it seems to me to continue to be rejected, rebuffed. I'm thrilled by the direction (and directives) of the current Chief of the Forest Service, Michael Dombeck; but he's a long damn way from Montana and the Yaak, and in the meantime, even though some wonderful and open Forest Service employees work up here, I still have grievances with some of the old guard.

In the Pine-Buckhorn sale, the Forest Service said they would helicopter log a roadless area and thin only six- to eight-inch Doug fir, to reduce fuel build-up. But when environmentalists went in to check, we found those smaller trees were left alone, with only the giants—the ones that would have survived a fire—scheduled to be cut. Upon being discovered in this position, the Forest Service decided upon what they thought was a compromise for this roadless area: to leave the giants *and* the flammable small trees, but just to whack the intermediate to large trees, down to a size-limit of twenty inches.

This isn't old-school stuff, tales from the woebegone past. This is 1999.

Such actions do not build trust.

At the Wood-Rat field trip, in the Yaak, we were told big larch trees would not be a primary component of proposed cutting units. In Unit 10 and other units in the Wood-Rat sale, big larch trees are the primary component.

This is not an old tale, either, but another 1999 story.

Such actions do not build trust.

At Clay-Beaver, in the Yaak, cutting units marched all the way to the roadless area's boundary and, in my opinion, carried over into it, with a clearcut scheduled at the edge of an area that is being considered for wilderness designation.

This is not an old story. This is a 1999 story.

Such actions do not build trust.

A handful of stressed-out volunteers visit what units we can in our valley. We can't get to them all; we provide our comments where we can, and, often, it feels as if our comments are marched over. *Dreams*, the administrators tell us, *you're dreaming.*

When we tell them we want peace, and a middle way, they say that we don't; they tell us that because we're critical, we're being difficult, refusing to go along. They threaten to stop providing us with information, threaten to shut down any dialogue, and complain in public about the environmentalists' "extremism."

Maybe I'm too close to it, emotionally—as close as you can get, really —but most days it feels like the only thing that's extreme is my despair.

I know time is going to heal the landscape here, and the community, and our dreams, my dreams, as it is doing in Vermont. I know too that I need to learn patience, and take a more glacial view. It's just that for such a long time, you feel like you're so close. As if wildness—protected, assured wildness—is only a step away.

One more step. Then one more. Then another. Just one more. You push on.

All this talk about a theoretical middle way, and all these lamentations about how difficult it is, in this age of craziness, to hold on to such a moderate vision: You can keep asking for the middle until one day you look up and see that not only has the middle been eaten away, but also the new middle that came after that, and so too the new middle that came after that one, and the one that followed that, until, once again, you find yourself at the precipice.

I need to hold onto *this* middle—the one I have been speaking of— fiercely; to not allow this middle, this last middle, to be eaten. I continue to believe that both ways, both needs, can still barely be achieved, and that this double-attempt, double-work, is the right thing to do.

That said, I fully reserve the right to change my mind. To believe, a year from now, or two years, that it's all a crock.

But for now, barely, I can still see a way. Barely, and if we hurry.

Where are the wolves? The glacier is moving fast now, slipping; if the wolves hurry, they can travel along its edges, weaving through and over the scattered moraine of its leavings. The glacier is moving, but also melting, deflating like the husk of some once-bright and living thing.

A new world is being made quickly now, a new landscape that is all but unrecognizable to the past—and if they hurry, wolves might find a place in it.

The great New England ice storm of 1998, for example, turned huge swaths of forest upside down. Classic "disturbance ecology," millions of acres of "disturbance," converted, in a single storm, shady late-successional forests to the more early seral stages preferred by deer: shafts of light crisscrossed randomly through the forest.

As well, the cores, the ecological reserves—hard-fought-for, hard-earned—are being secured, through purchases and conservation easements. A tattered fabric is being repaired, woven back together: not complete, but perhaps firm enough to hold the resurrected passings of grace.

Such changes might bode well for the wolves, or for the idea of wolves that is skirting the edges of our present history—wolves presently traveling along the outer arc of these changes as if trying to decide whether to re-enter.

Foresters in the West sometimes speak of a stand of trees becoming "decadent," and "falling apart," implying that the cumulative stresses of the world and time have become too great. It is as if, finally, one too many threads has been plucked from the fabric of the whole; that an unraveling has begun, and that the momentum of that falling-apart has become irreversible, that the collapse must play itself all the way out to the end, before things can be turned back around again, and "growth," or "progress," achieved, or re-achieved.

Nearly anyone who has spent time in the woods is able to notice with alarming intensity the changes hurtling at us—the going away of specific and crafted things, and the burgeoning of the abstract, the general, the accommodating, and the compromised.

All over the eastern forests, red maples are overtaking the mast-producing oaks and hickories that deer, wild turkeys, and squirrels rely upon—indeed, trees that the ecology of the entire forest relies upon. The oaks and hickories have been present in the East for ten thousand years, since the fading-away of the last ice, but now they are tumbling, as if in another kind of overnight storm. They are succumbing to the ravages of the gypsy moth. A walk in the woods where these defoliants are active by the millions is unmistakable; you can scent the caterpillars themselves, and the tannic odor of the green leaves being chewed; and you can hear the pitter-pat, like rain, of countless caterpillars shitting, defecating their "frass" to the forest floor, as the forest of an eon disappears—as if it were only a dream or a story, a story written in pencil, a story now being erased.

Farther south, in the Appalachians, dogwoods are falling to pandemic fungal outbreaks for which no tree appears to have resistance. No one knows where the fungus has come from, and scientists are amazed that in just a few years it has spread throughout the dogwood's range in the East. Newly tipped meteorological imbalances? Acid rain? The import of foreign stock? No one knows—only that the dogwoods, as Charles Little chronicles in *The Dying of the Trees*, are indeed dying, as did the elm and chestnut in the East before them.

Vanishing too in the East are the sugar maples; dramatically so. They are simply not regenerating. A study by the Wildlife Conservation Society has found reversed demographic curves in the Adirondacks—not by a few percentage points, but by a factor of twenty, in some regions.

Something vital, integral, has collapsed, some balance, and foresters aren't sure why, or what it is. Over-browsing by deer might be a factor in some places, as might be the suppression of fire, or past logging practices, but none of these suggestions or hypotheses can be applied across the spectrum of the sugar maple's distribution. The failure to regenerate is total, so that it is as if the problem lies in the soil itself—which points, as do so many of the East's ecological woes, directly to acid rain.

There aren't many new, young sugar maples coming "on line." When the old ones die, that might be the end of that story, or that chapter in the story.

Dying too are the red spruce, in great sheets and waves, probably from the drift of skyborne acid—dying hardest and most often at the higher elevations. And even in places where spruce (and nearly every other major species) are not dying, they are in decline, exhibiting symptoms that include "very slow growth, early leaf drop, snap-off, root decay, discolored foliage, and premature death."

Such decadence, in proportion, is healthy, even luxurious, in any forest. Charles Little, and others, however, report a pandemic: a shop of ecological horrors.

Insect infestations, previously cyclic and pulse-like, are now sweeping over the dead and dying forests not like a tide, which would carry some regress with each egress, but like a flood that keeps only rising. Black flies thrive in acidic waters. Woolly adelgids—nitrogen lovers—are destroying hemlock and other conifers. Asian long-horned beetles (ALB) and other insects are "hitchhiking" to every part of our country on untreated pallets and packing crates; the ALB is particularly dangerous for wildlife, as it targets the hardwood species so important to wildlife: maples, poplars, and aspen. Bark beetles, caseworm bearers, Tussock moths, and dozens of others are combining to deliver a plague to the country's forests that is biblical in its proportions. (In Germany, there is even a name in their culture for this kind of forest death: *Waldsterben*.)

I want wilderness. I want to believe in—to know of—dark wild sanctuaries of untouched, *unmanipulated*, grace and mystery, power and health.

Undeniably, those larger forests that we have not yet built roads into, and that we have not yet altered with logging practices (even well-meaning or "sustainable" practices) are the healthiest. We should immediately take steps to protect these last untouched forests—the last of the last. Even across the span of the last few years, I have noticed in my own valley the blazing-quick spread of invasive weeds—knapweed and thistle first, and now the far-more dangerous hawkweed—robbing the meadows of the grass that the elk rely upon, scorching the forest floor as if with a kind of sterile fire. In the last couple of years, too, I've noticed the silkworm moths' invasion, following, like the weeds, the route of roads, farther and farther into the forest. August is a hard month for the soul up here, when all the weeds are in full bloom. Even ten years ago, these weeds were barely noticeable, or not present at all. We didn't even have ticks. It was as close to Eden as anything I've seen in the Lower Forty-Eight.

Now, in Montana—as in Vermont—we are simply trying to gather and hold together the larger more vital pieces, like museum exhibits, or like islands within islands within islands.

What if you love salamanders as much as you love wolves? How then to hold the statistic that clearcutting kills an estimated 14 million salamanders per year, by robbing the dense shady moisture-trapping overstory that they rely upon to exist in their damp protected microhabitats?

What irony, then, if we finally recover wolves to the Northeast, but lose damn near everything else.

We need more wilderness.

Have I been too indulgent in utilizing this opportunity to speak about the nonexistent Northeastern wolf to instead wander two thousand miles west and spend so much ink upon a dream of "good" logging, and the value of rural communities, and our last wild, healthy, untouched road-less areas, and to praise the patience of glaciers?

To speak, in the absence of wolves, of the wolves' habitat: the deep forest, and the landscape of human politics?

If I have, then an apology is due not to the reader, but to wolves, and to the idea of wolves. In my valley, where wolves are occasionally present, passing through more like dreams or fog, I have seen their tracks and scat fairly frequently, though I have only glimpsed actual wolves on four separate occasions in fourteen years, and have only heard them a very few times.

Nonetheless, it is a source of great balm and confidence to me to know that they are here. They help anchor my emotional and interior landscape to the physical landscape. I love the fact that they are rarely, if ever, seen. I love that they have gone away and now are coming back.

I love too that it often feels we are undeserving of them, and yet here they come: the idea of them, first, but soon enough, the shadows, and then the desire, and then the wolves themselves.

Where are the wolves, in this story?

I guess that answer is up to us: to first make a place that is ready for them. In our hearts, surely—but harder still, out on the land.